From Ice to Inferno

What Earth's Past Reveals About Our Future

Douglas B Sims, PhD

Douglas B Sims, PhD

From Ice to Inferno

For more information, or to book an event, contact:
dsims@simsassociates.net

Book design by DB Sims
Cover picture purchased from: iStock (RapidEye)

ISBN – Paperback: 978-1-966739-08-1
ISBN – eBook: 978-1-966739-09-8

First Edition: April 2025

Please leave a review on Amazon, Goodreads, or any site that you purchased this book as a review is part of the overall experience.

Douglas B Sims, PhD

Table of Contents

Douglas B Sims, PhD

Acknowledgements

I am deeply thankful to my wife, whose unwavering support, wisdom, and love have been my constant foundation and source of inspiration. The 34 years we've shared have enriched not only my life but also this project. Your encouragement has lifted me through every challenge, and your insights are woven into every page.

To our children, thank you for bringing joy and growth into our lives and teaching us the profound lessons of parenthood. Watching you grow into the people you are today has been one of my greatest privileges, filling me with pride and shaping my perspective in countless ways.

To my friends and colleagues, particularly those in paleoclimate studies and political science, thank you for your invaluable insights and perspectives. Our conversations and debates have enriched my thinking and brought depth to this book that I could not have achieved on my own.

Finally, to the many professionals and individuals I've had the honor to learn from, thank you for sharing your stories and experiences. Your voices have provided real-world context and understanding that resonate throughout these pages.

Douglas B Sims, PhD

Introduction

Earth's climate has always been a story of change, a slow, sweeping epic shaped by the interplay of atmosphere, oceans, land, and life. For billions of years, these shifts unfolded gradually, giving ecosystems and species time to adapt and evolve.

But today, we stand at a crossroads unlike any before. Human activity has accelerated the pace of climate change to levels never before seen in Earth's long history. What once took millennia is now happening in mere decades.

This book is a chronicle of our planet's climatic past and a call to action for its future.

It takes you on a sweeping journey through the ice ages, the interglacials, the fragile window of stability that allowed civilization to flourish and now, the tipping points we can no longer ignore.

From the retreat of ancient ice sheets to the stability of the Holocene, humanity has lived in a rare moment of climatic fortune. But with that fortune comes responsibility. Industrialization has pushed atmospheric greenhouse gases to dangerous levels, triggering a cascade of environmental challenges: rising seas, extreme weather, biodiversity collapse.

The strength of From Ice to Heat lies in connecting the past to the present—and showing why this moment demands more from us than ever before.

History teaches us that Earth can survive great transformations. The question now is: Can we?

This is not just a scientific account. It is a deeply human story—one that asks:

- Can we halt the momentum of climate change before it reshapes our world beyond recognition?

- Can we adapt to the transformations already underway?

- Can we come together, across borders and generations, to protect the only home we have?

The answers are daunting, but they are not hopeless. Humanity's history is rich with innovation, resilience, and cooperation in the face of overwhelming odds. That same spirit must guide us now.

As you read, remember: the future is not written yet. The next chapter of Earth's climate story depends on us, our choices, our courage, and our commitment to change.

The time to act isn't tomorrow.

It's right now.

Chapter 1

Earth's Climate Odyssey

Earth's climate history is a tale of dramatic shifts and cyclical transformations, driven by natural forces over billions of years. From the frozen expanses of ancient glaciations to the warmer interglacial periods that fostered the rise of civilizations, these climatic changes have shaped the planet's landscapes, ecosystems, and the course of human development. Today, as humanity faces unprecedented challenges from anthropogenic climate change, understanding this intricate past offers critical insights into the forces that govern our environment and the pathways to a sustainable future. This chapter explores the major epochs of Earth's climate history, the significance of glaciations and interglacial periods, and humanity's evolving role in influencing the planet's climate.

Overview of Earth's Climate History

Understanding Earth's climatic past is not an academic exercise; it is a practical necessity for planning humanity's future. Earth's climate history is a complex narrative of dynamic changes, marked by cycles of glaciation, interglacial warming, and dramatic shifts in atmospheric composition. Over the past 4.6 billion years, the planet has transitioned through various climatic epochs, each shaped by intricate interactions between natural forces such as volcanic activity, orbital variations, solar radiation, and plate tectonics (Ruddiman, 2014). These transitions have profoundly influenced the evolution of life and the development of ecosystems, underscoring the critical role of climate in shaping Earth's geological and biological history.

One of the most dramatic episodes in Earth's climate history occurred during the Proterozoic Eon (2.5 billion to 541 million years ago), which witnessed some of the earliest known ice ages. Evidence from geological formations suggests that Earth may have experienced periods of near-total glaciation, known as "Snowball Earth" events, when the planet's surface was almost entirely frozen (Kirschvink, 1992). These episodes were likely triggered by a combination of reduced greenhouse gas concentrations and a decrease in solar luminosity, compounded by feedback mechanisms such as increased albedo from extensive ice cover. The effects of these glaciations were profound, reshaping Earth's surface, creating new ecological niches, and setting the stage for the evolution of multicellular life during the subsequent Cambrian explosion.

In more recent geological history, the Quaternary Period (2.58 million years ago to present) has been characterized by cyclical glacial and interglacial periods. These cycles are driven primarily by Milankovitch cycles—variations in Earth's orbital eccentricity, axial tilt, and precession—that affect the distribution of solar radiation on the planet's surface (Hays, Imbrie, & Shackleton, 1976). During glacial periods, massive ice sheets extended over large portions of North America, Europe, and Asia, profoundly altering global sea levels, ecosystems, and atmospheric circulation patterns. For instance, the Last Glacial Maximum (LGM), which peaked approximately 20,000 years ago, saw sea levels drop by about 120 meters, exposing continental shelves and land bridges, such as the Bering Land Bridge, which facilitated the migration of humans and other species (Clark et al., 2009).

The end of the LGM marked the transition into the Holocene Epoch, an interglacial period that began around 11,700 years ago. The Holocene has been a time of relative climatic stability, with warmer global temperatures allowing for the development of agriculture, the rise of complex societies, and the rapid expansion of human civilizations. However, this stability has also been punctuated by smaller climatic events, such as the Younger Dryas, a brief return to near-glacial conditions approximately 12,900 years ago, likely caused by disruptions in ocean circulation patterns.

Throughout Earth's history, climate has been regulated by intricate feedback mechanisms that maintain the delicate balance of the climate system. For instance, changes in albedo—the reflectivity of Earth's surface—play a significant role in amplifying or mitigating temperature changes. During glacial periods, extensive ice cover increases albedo, reflecting more solar radiation and reinforcing cooling. Conversely, during interglacial periods, ice retreat lowers albedo, allowing more solar energy to be absorbed and promoting warming. Additionally, the concentration of greenhouse gases such as carbon dioxide (CO_2) and methane (CH_4) in the atmosphere has been a critical factor in regulating global temperatures. Natural processes, including volcanic eruptions, weathering of rocks, and biological activity, have historically governed these gas levels, but human activities now significantly influence their concentrations.

Ocean circulation patterns, particularly the thermohaline circulation, also play a pivotal role in regulating climate by redistributing heat and regulating carbon storage. Disruptions to these patterns, whether from natural events or anthropogenic influences, can trigger abrupt climate shifts with cascading effects on ecosystems and weather patterns.

This intricate interplay of natural forces highlights both the resilience and fragility of Earth's climate system. While the planet has demonstrated a remarkable ability to recover from past climatic extremes, the unprecedented rate of current anthropogenic changes poses significant challenges. Understanding the processes that have shaped Earth's climate over billions of years is critical for predicting future trends and mitigating the impacts of human-induced climate change. This knowledge not only helps us contextualize our current climate challenges but also equips us with the tools to navigate a path toward a sustainable future.

Importance of Understanding Glaciations, Interglacial Periods, and Climate Shifts

Comprehending the dynamics of glaciations and interglacial periods is essential for understanding the mechanisms that govern Earth's climate. Glacial periods, characterized by the extensive growth of ice sheets and glaciers, occur when reduced solar radiation reaches the Earth's surface due to changes in the planet's orbital configuration—variations collectively known as Milankovitch cycles (Hays, Imbrie, & Shackleton, 1976). These cycles, which include fluctuations in Earth's eccentricity, axial tilt, and precession, influence the distribution and intensity of solar energy, creating conditions that either promote ice sheet expansion or foster melting. In contrast, interglacial periods represent warmer phases in the cycle, marked by the retreat of ice, stabilization of global temperatures, and the flourishing of ecosystems.

The shifts between these phases have profound implications for global sea levels, ecosystems, and biodiversity. During glacial periods, vast amounts of water are stored in ice sheets, causing sea levels to plummet. For instance, during the Last Glacial Maximum (LGM), approximately 20,000 years ago, global sea levels were about 120 meters lower than today, exposing extensive areas of continental shelf and creating land bridges like the Bering Land Bridge. This bridge was a critical migration pathway, enabling early humans and other species to move between Asia and North America (Lambeck et al., 2014). Such shifts not only reshaped the planet's geography but also influenced evolutionary pathways, driving species to adapt to dramatically altered habitats.

Interglacial periods, by contrast, witness the melting of ice sheets, leading to rising sea levels and expanded marine habitats. These warmer phases have historically provided stable climates that support the growth of biodiversity and the establishment of human civilizations. However, they are not without variability. Sudden climatic events, such as the Younger Dryas, demonstrate how sensitive these periods can be to changes in ocean circulation or other feedback mechanisms.

The study of glaciations and interglacials is invaluable for predicting future climate scenarios. By examining paleoclimate data from sources like ice cores, tree rings, and sediment layers, scientists can reconstruct past atmospheric conditions, including greenhouse gas concentrations and temperature fluctuations. For example, ice cores from Antarctica and Greenland contain trapped air bubbles that preserve a record of atmospheric CO_2 levels over hundreds of thousands of years, providing direct evidence of how these gases correlate with temperature changes (Petit et al., 1999). Tree rings, meanwhile, offer annual records of growth patterns influenced by temperature and precipitation, while sediment layers reveal long-term changes in vegetation and water chemistry.

This wealth of data allows scientists to identify patterns and feedbacks that have historically driven climate transitions, offering insights into how current anthropogenic influences might alter these dynamics. The rapid increase in greenhouse gas concentrations since the Industrial Revolution—far exceeding the levels observed in previous interglacial periods—suggests that humanity has become a significant driver of climate change. Understanding the natural rhythms of glaciations and interglacials provides a crucial baseline against which to measure and predict the trajectory of human-induced changes.

These insights are more than academic; they hold practical implications for mitigation and adaptation strategies. Predicting sea-level rise, for example, depends on understanding how ice sheets respond to warming, while forecasting shifts in ecosystems requires knowledge of how species have historically adapted to climate transitions. By studying the glacial and interglacial dynamics of the past, we gain a deeper appreciation for the resilience and fragility of Earth's climate system, equipping us with the tools to navigate a rapidly changing future.

Events Shape the Environment and Humanity's Role in Current and Future Changes

Earth's glacial and interglacial cycles have left profound and lasting impacts on the environment, influencing not only physical landscapes but also ecosystems and human history. During glacial periods, massive ice sheets carved the Earth's surface, creating valleys, fjords, and other dramatic landforms. These glaciers deposited nutrient-rich soils in regions such as the Great Plains, laying the foundation for fertile agricultural lands that have sustained human societies for millennia (Ehlers & Gibbard, 2007). Interglacial periods, in contrast, were characterized by the retreat of ice and the expansion of forests, grasslands, and other ecosystems, enabling biodiversity to thrive in newly available habitats. Species adapted to the cooler conditions of glaciations often had to migrate or evolve rapidly during interglacials, demonstrating the dynamic interplay between climate and evolution.

Humanity's role in the current climate trajectory, however, marks a significant departure from these natural cycles. Industrialization has introduced a new and unprecedented factor into the climate equation, with human activities significantly amplifying natural warming processes. Atmospheric carbon dioxide (CO_2) levels, which historically remained below 300 parts per million (ppm) during interglacial periods, have now exceeded 420 ppm—a dramatic increase driven by fossil fuel combustion, deforestation, and other industrial practices (NOAA, 2023). This anthropogenic acceleration of greenhouse gas emissions is disrupting Earth's natural climatic rhythms, pushing the planet into uncharted territory.

The concept of the Anthropocene—a proposed epoch defined by humanity's dominant influence on Earth's systems—underscores the scale and urgency of these changes. Climate models warn that without immediate and effective mitigation efforts, global temperatures could rise by 2–4°C by the end of the 21st century. Such an increase would have catastrophic consequences, including significant sea-level rise, more frequent and severe extreme weather events, and widespread biodiversity loss. The Intergovernmental Panel on Climate Change (IPCC) projects that these changes could destabilize ecosystems, displace millions of people, and pose existential threats to human and natural systems alike (IPCC, 2021).

Despite the gravity of the challenges, humanity also holds an extraordinary capacity to influence positive change. Advances in renewable energy, innovations in carbon capture and storage, and international efforts such as the Paris Agreement demonstrate the potential for collective action to combat climate change. Learning from the resilience strategies of the past—both natural and human— provides valuable lessons for adaptation and mitigation. For example, understanding how ecosystems rebounded after glacial retreats can inform strategies for restoring degraded environments and conserving biodiversity.

The knowledge gained from studying Earth's glacial and interglacial cycles also offers hope for navigating the future. These cycles reveal the planet's remarkable resilience and its ability to recover from climatic extremes. While the current pace of change is unprecedented, humanity's ability to innovate and cooperate has the potential to steer the planet toward a sustainable future. By harnessing scientific understanding and embracing global cooperation, we can rise to meet the challenges of the Anthropocene, ensuring that future generations inherit a planet capable of supporting life in all its diversity.

The story of Earth's climate is one of resilience and transformation, shaped by powerful natural forces and, more recently, human actions. As we stand at a crossroads, the lessons of the past remind us that while climate shifts are inevitable, their severity and impact are not beyond our influence. Humanity has the tools, knowledge, and capacity to mitigate the damage, adapt to change, and forge a sustainable future. By understanding the cycles that have shaped our world and recognizing the urgency of our role, we can become stewards of a planet that thrives not just for our generation but for all those to come. The history of Earth's climate is a call to action, urging us to align our ingenuity with the enduring rhythms of nature.

As we transition to Chapter 2, we shift our focus from Earth's climatic narrative to humanity's scientific quest to understand it. From ancient observations to advanced models, the story of climate science reflects both our growing awareness and our expanding impact. If Chapter 1 told the story of Earth's climate, Chapter 2 begins the story of how we came to know it—and what we now must do with that knowledge.

Chapter 2

Unlocking Earth's Climate Code

The science of climate change has evolved significantly over the past centuries, from early theories about the Earth's climate to today's sophisticated models predicting future impacts. Understanding how climate has changed over geological time provides essential context for today's challenges, revealing both natural cycles of warming and cooling and the unprecedented rate of change driven by human activity. This chapter explores the history of climate science, from the early observations of greenhouse gases to the development of modern climate models, and examines the key historical shifts in the Earth's climate, such as glaciations and interglacial periods, to highlight how these past changes are relevant to current climate concerns. It also addresses the profound impact of industrialization, particularly the rise of fossil fuel consumption and deforestation, and how these activities have altered the balance of Earth's climate, setting the stage for the climate crisis we face today.

Evolution of Climate Science and Historical Climate Change

Understanding the science behind climate change requires not only a deep dive into the historical fluctuations of Earth's climate but also a journey through the evolution of scientific thought that has shaped our current understanding. Early ideas about climate were largely based on empirical observations, with early thinkers noticing seasonal patterns and the effects of the Sun's cycles on the Earth's climate. Ancient civilizations, for example, recorded patterns of seasons and weather to guide agriculture, but the formal scientific study of climate didn't emerge until much later.

The first attempts to scientifically understand Earth's climate began with early philosophers and astronomers who speculated about the relationship between the Earth and the Sun. Ancient Greeks such as Aristotle recognized the variability of the Earth's climate, but it was not until the 17th and 18th centuries, during the Age of Enlightenment, that scientific thought about climate began to emerge more systematically. Early theories were based on simple observations of weather and seasonal changes, and scientists such as Galileo and Johannes Kepler began to explore the role of solar radiation and planetary motions in affecting the Earth's climate.

The modern understanding of climate change, however, began to take shape in the 19th century, when a series of breakthroughs in science laid the foundation for today's climate science. One of the most pivotal moments in this evolution was the development of the greenhouse effect theory. In 1824, French mathematician Joseph Fourier proposed the idea that the Earth's atmosphere acted as a "blanket," trapping heat from the Sun. Fourier's hypothesis was based on his observations that the Earth was warmer than it would be if it simply reflected the Sun's heat back into space, suggesting that the atmosphere played a crucial role in warming the planet (Fourier, 1824).

Building on this, Swedish chemist Svante Arrhenius, in 1896, proposed that increases in atmospheric carbon dioxide (CO_2) from human activities, particularly the burning of fossil fuels, could lead to global warming. This marked one of the earliest scientific recognitions of human influence on climate change. Arrhenius' work showed that CO_2 acted as a greenhouse gas, trapping heat in the atmosphere, and he estimated that a doubling of atmospheric CO_2 could increase Earth's temperature by around 5-6°C—an insight that laid the groundwork for modern climate models (Arrhenius, 1896).

In the 20th century, climate science advanced dramatically with the development of more sophisticated models and tools. The advent of computers in the 1950s allowed scientists to develop the first climate models that could simulate the interactions between the atmosphere, oceans, and landmasses. One of the most important early models was created by Charles David Keeling, whose work in the 1950s led to the development of the Keeling Curve—a graph that shows the rise in atmospheric CO_2 measured at Mauna Loa, Hawaii. Keeling's work provided irrefutable evidence that CO_2 concentrations were steadily increasing, and this became one of the key indicators of human impact on the Earth's climate (Keeling, 1960).

By the 1970s, scientists had begun to recognize that the increase in greenhouse gases was not only a natural process but also the result of industrialization and human activities. The rise of environmentalism and the establishment of institutions like the Intergovernmental Panel on Climate Change (IPCC) in 1988 helped further solidify climate change as a global scientific concern. These efforts culminated in a broad consensus among scientists that human activities were indeed contributing to global warming, and this would have profound effects on weather patterns, sea levels, and ecosystems.

Looking back at the Earth's historical climate shifts, scientists have been able to use proxies such as ice cores, tree rings, and sediment layers to reconstruct past climate data. These records show that Earth's climate has undergone significant fluctuations, from ice ages to warm interglacial periods. One of the most famous historical climate shifts is the Ice Ages, during which large portions of the Earth were covered in thick ice sheets. These ice ages were driven by natural factors like variations in Earth's orbit (the Milankovitch cycles), but the current phase of rapid warming is a stark departure from these natural cycles.

The modern understanding of climate change now rests on the knowledge that both natural and human factors contribute to the Earth's climate system. While natural processes such as volcanic eruptions, solar variability, and ocean currents play a role in climate fluctuations, it is clear that human activities—particularly the burning of fossil fuels, deforestation, and industrial processes—are the dominant drivers of the current period of rapid climate change. Unlike the gradual changes of past climate shifts, the current pace of warming is unprecedented, and it is this rapid, human-driven change that poses the greatest threat to ecosystems, economies, and human societies.

This evolving understanding of climate science has shaped the ways we view and respond to climate change. From early theories about the role of the atmosphere in warming to the sophisticated climate models used today, our scientific understanding has advanced significantly. However, as we now know, the science of climate change is not just about understanding the past and the present—it's also about predicting future trends, understanding the potential impacts of those trends, and developing solutions to mitigate the damage already done. The history of climate science is a testament to the ability of human thought and innovation to expand our understanding, but it also highlights the urgency of acting on that knowledge to address the climate crisis before it becomes irreversible.

Early Theories and the Beginnings of Climate Science

One of the first recorded attempts to explain the Earth's climate came in the 19th century when scientists began to understand the atmosphere's role in regulating temperature. This marked a pivotal shift in scientific thinking, as researchers moved from purely observational studies to formulating hypotheses about the Earth's energy balance. The most notable of these early attempts came from French mathematician Joseph Fourier, who, in 1824, proposed the revolutionary idea that the Earth's atmosphere acts as a "blanket" that traps heat. Fourier's hypothesis was grounded in the observation that Earth was warmer than it would be if it simply reflected the Sun's heat back into space. He theorized that the Earth's atmosphere retained heat from the Sun through a process akin to the way a blanket traps warmth around the body. Fourier's work laid the foundation for future studies of what we now call the greenhouse effect—a process where certain gases in the atmosphere, particularly water vapor and carbon dioxide (CO_2), trap heat and prevent it from escaping into space. This concept of the atmosphere as a heat-trapping "blanket" fundamentally altered the way scientists approached the Earth's climate system and set the stage for further advancements in climate science (Fourier, 1824).

Fourier's work was groundbreaking, but it remained largely theoretical until the late 19th century, when Swedish chemist Svante Arrhenius made a significant contribution to climate science that would shape the field for generations. In 1896, Arrhenius proposed that increases in atmospheric carbon dioxide (CO_2) from human activities, particularly the burning of fossil fuels, could enhance the greenhouse effect and lead to global warming. At the time, this was a bold hypothesis, as the industrial revolution was in full swing, and scientists had not yet fully understood the extent to which human activity could influence the Earth's climate. Arrhenius's work demonstrated the link between CO_2 concentrations and temperature, suggesting that the accumulation of CO_2 in the atmosphere could cause a rise in global temperatures. He calculated that a doubling of atmospheric CO_2 could increase the Earth's temperature by 5 to 6°C, a figure that, while later refined, remains a central concept in modern climate science (Arrhenius, 1896).

Arrhenius's work was groundbreaking not only because it linked human activity to climate change but also because it was one of the first scientific studies to suggest that the Earth's climate could be influenced by factors beyond natural variability. This laid the groundwork for understanding the human influence on climate, a concept that has become a cornerstone of contemporary climate science. His pioneering studies highlighted the importance of CO_2 in regulating Earth's temperature and underscored the potential for anthropogenic emissions to alter the planet's climate.

As the 20th century progressed, scientific research into climate change became increasingly sophisticated, thanks to advancements in both computational technology and the growth of multidisciplinary research. During the mid-20th century, scientists began developing complex climate models capable of simulating atmospheric and oceanic conditions. These models, which used mathematical equations to represent the physical processes that govern Earth's climate, were initially used to understand local weather patterns but eventually evolved into tools capable of predicting broader climate trends. By the 1950s and 1960s, researchers had access to more powerful computers, allowing them to create more accurate and detailed climate models that could simulate long-term changes in temperature, precipitation patterns, and weather systems.

One of the most important breakthroughs in climate modeling came with the development of global climate models (GCMs), which integrated data from a variety of sources, including atmospheric temperatures, CO_2 concentrations, and ocean currents, to predict long-term climate trends. These models helped scientists understand how rising concentrations of greenhouse gases, particularly CO_2 and methane (CH_4), would affect global temperatures and weather systems. By the 1970s, the scientific community reached a broad consensus that human activities, particularly the burning of fossil fuels and deforestation, were contributing to the accumulation of greenhouse gases in the atmosphere. This consensus was bolstered by the work of climate scientists such as James Hansen, whose research in the 1980s warned of the potential dangers of unchecked global warming. Hansen's testimony before the U.S. Congress in 1988 was a pivotal moment in climate science, bringing public attention to the growing evidence of human-caused climate change.

By the late 20th century, the science of climate change had advanced significantly, with the establishment of international bodies such as the

Intergovernmental Panel on Climate Change (IPCC) in 1988 to coordinate research and provide policy recommendations. The IPCC's reports, based on the work of thousands of scientists worldwide, have played a crucial role in advancing our understanding of climate change and its potential impacts. These reports have provided comprehensive assessments of the state of the climate, predicting the consequences of rising greenhouse gas concentrations on global temperatures, sea levels, and ecosystems.

Today, climate science has evolved into a highly interdisciplinary field that draws on expertise from meteorology, oceanography, physics, chemistry, biology, and economics. With the help of sophisticated models, observational data from satellites and ground-based monitoring stations, and ongoing research, scientists continue to refine our understanding of climate change and predict future scenarios. However, the fundamental concepts introduced by Fourier, Arrhenius, and others—such as the role of greenhouse gases in regulating Earth's temperature—remain as relevant today as they were more than a century ago. The rapid warming we are experiencing now, driven primarily by human activities, serves as a stark reminder of how far our understanding of climate has come and how much more urgent the need for action has become.

Glaciations and Interglacial Periods

To fully grasp the contemporary climate crisis, it is essential to consider the Earth's historical climate shifts, especially the cycles of glaciation and interglacial periods. Over the past 2.6 million years, known as the Quaternary period, the Earth has undergone a series of ice ages, which have profoundly shaped its climate. These ice ages, characterized by the expansion of vast ice sheets across large parts of the Northern Hemisphere, were punctuated by warmer interglacial periods, like the current epoch, the Holocene, which began approximately 11,700 years ago. Understanding these natural climate cycles is crucial for contextualizing the current climate crisis and recognizing the profound changes that human activity has introduced into a system that has, until now, evolved on much slower timescales.

The Pleistocene Epoch and the Last Glacial Maximum (LGM)

The most recent ice age, the Pleistocene epoch, was marked by the cyclical growth and retreat of massive ice sheets. During periods of glaciation, ice covered vast portions of the Earth's surface, particularly in the Northern Hemisphere, where enormous ice sheets extended from the polar regions to as far south as the northern United States, parts of Europe, and northern Asia. The Earth's climate during these periods was drastically cooler, and sea levels were much lower due to the vast amounts of water locked in ice.

The Last Glacial Maximum (LGM), around 20,000 years ago, represented the peak of the most recent glaciation, when ice sheets reached their greatest extent. During this period, much of North America, Europe, and Asia were covered in thick ice sheets, with the Laurentide Ice Sheet in North America being particularly dominant. At its peak, the ice sheets were up to several kilometers thick, shaping landscapes through processes of erosion and deposition. For instance, the formation of the Great Lakes in North America and the fjords of Scandinavia can be traced back to the movement of these massive ice sheets.

As the climate began to warm around 15,000 years ago, these ice sheets began to retreat, marking the transition into the present interglacial period. This warming was not uniform, and the rate of temperature increase fluctuated, but by about 11,700 years ago, the Earth had entered the Holocene, a relatively stable and warmer climate phase, which has supported the rise of human civilization.

The Natural Drivers of Climate Change

The natural cycling between glaciations and interglacials has been driven by variations in Earth's orbit and axial tilt, a concept known as the Milankovitch cycles, named after the Serbian scientist Milutin

Milankovitch who first described them in 1941. These cycles include three main components: eccentricity (the shape of Earth's orbit), axial tilt (the angle at which Earth is tilted on its axis), and precession (the wobble of Earth's axis). Together, these variations affect the amount and distribution of solar radiation received by Earth, influencing the timing and intensity of ice ages.

The Earth's orbit oscillates from being more circular to more elliptical over a period of approximately 100,000 years (eccentricity), the tilt of its axis varies between 22.1° and 24.5° on a 41,000-year cycle, and the axis of rotation wobbles over a 26,000-year period. These cycles interact in complex ways to alter the seasonal distribution of sunlight, particularly at high latitudes, which in turn affects the growth and retreat of ice sheets. The Milankovitch cycles are therefore considered the primary natural drivers of the long-term climate changes that have shaped Earth's history.

The Speed of Current Warming

While these natural cycles have caused significant fluctuations in the Earth's climate, the current rate of warming is unlike anything seen in the past. In contrast to the gradual temperature shifts that occurred over tens of thousands or even hundreds of thousands of years during past interglacial periods, modern climate change is happening at an alarming pace. Global temperatures have already increased by approximately 1.1°C since the late 19th century, and this warming is primarily attributed to human activities, particularly the burning of fossil fuels such as coal, oil, and natural gas, which release carbon dioxide (CO_2) and other greenhouse gases into the atmosphere.

The rapidity of this warming is far outpacing any natural fluctuations in the Earth's climate history. For instance, during the transition from the last glaciation to the Holocene, the Earth's temperature rose by roughly 5-6°C over thousands of years. Today, human-induced warming is projected to surpass that threshold by the end of this century if current emissions trends continue (IPCC, 2021). The speed at which global temperatures are rising—driven by the accumulation of CO_2, methane, and nitrous oxide—has never been seen before in the context of natural climate cycles.

A Departure from Natural Cycles

Human activities have radically altered the Earth's natural climate dynamics, making the current situation unprecedented in terms of the scale and speed of change. The burning of fossil fuels, deforestation, and land-use changes have led to an accumulation of greenhouse gases in the atmosphere, trapping more heat and intensifying the greenhouse effect. This human-driven alteration of the climate system has overwhelmed the natural processes that once governed the Earth's climate. While the Milankovitch cycles still influence the planet's climate over long timescales, the intensity and rapid pace of the current warming are primarily due to human influence, rather than natural variability.

In addition to the effects of rising greenhouse gas concentrations, human activities have introduced new variables into the climate system, including land use changes and deforestation, which exacerbate the impacts of climate change. Deforestation, for example, reduces the planet's ability to absorb CO_2 through photosynthesis, further contributing to rising atmospheric carbon levels. Similarly, the urbanization and industrialization associated with human development have led to the release of pollutants that not only contribute to warming but also affect regional climate patterns, such as through urban heat islands or the alteration of local precipitation patterns.

The Urgency of Addressing Human-Driven Global Warming

The historical climate shifts we observe through past glaciations and interglacial periods are important because they highlight the rarity of rapid climate change in Earth's history. Natural climate cycles have caused shifts in temperature over millennia, but the current rate of warming, driven by human activities, is unprecedented. The changes we are experiencing today are occurring faster than the Earth has ever warmed or cooled in the past, and they are bringing about far-reaching consequences, including rising sea levels, more frequent and severe weather events, and disruptions to ecosystems.

The historical context of climate change underscores the urgency of addressing human-driven global warming. The natural variability in Earth's climate has been a constant force, but the rapid, anthropogenic warming we are experiencing today requires immediate and decisive action. Understanding the past cycles of warming and cooling helps us appreciate the importance of mitigating the pace of current climate change and underscores the need to shift away from the practices that have contributed to this accelerated warming. If humanity does not take immediate steps to reduce emissions and adapt to the changing climate, we risk exceeding critical thresholds that could lead to irreversible environmental and societal consequences.

Relevance to Today's Climate Crisis

The historical context of climate change is crucial for understanding the magnitude of the current crisis. Throughout Earth's history, the climate has undergone significant shifts, driven by natural factors such as volcanic activity, changes in solar radiation, and variations in the Earth's orbit. These shifts, however, have typically occurred over long periods, often spanning thousands or millions of years. What sets the current climate crisis apart is the unprecedented speed at which it is happening and the direct link to human activity, particularly the burning of fossil fuels and deforestation. This rapid pace of warming, combined with the extraordinary levels of carbon dioxide (CO_2) in the atmosphere, presents a stark contrast to past natural climate fluctuations and underscores the urgency of addressing the crisis.

Historically, climate shifts, such as glaciations and interglacial periods, have had profound impacts on ecosystems, sea levels, and even the rise and fall of human civilizations. During periods of glaciation, much of

Earth's surface was covered by thick ice sheets, leading to lower sea levels and drastically altered ecosystems. As the Earth warmed and entered interglacial periods, ice sheets retreated, sea levels rose, and ecosystems adapted to the changing conditions. These transitions were slow, taking thousands of years, and ecosystems had time to adjust. However, with the onset of human-driven climate change, the pace of warming is far more rapid, and ecosystems, human societies, and the planet as a whole are struggling to keep up.

The current trajectory of global warming threatens to induce disruptions that are potentially more severe than those seen in the past. Rising temperatures are expected to cause the melting of polar ice caps and glaciers, leading to rising sea levels. This will have profound effects on coastal communities, which are home to billions of people. The rapid increase in temperature also exacerbates the frequency and intensity of extreme weather events, such as hurricanes, heatwaves, and droughts, leading to significant impacts on agriculture, water resources, and human health. These changes will affect biodiversity, with many species unable to adapt quickly enough to survive in the face of rapidly changing habitats and climates. As ecosystems become destabilized, the impacts will cascade through food chains, leading to disruptions in global food security and the livelihoods of people who depend on these natural resources.

One of the most striking differences between the current climate change trajectory and past climate shifts is the role of human activity. Natural cycles, such as the Milankovitch cycles, have caused slow fluctuations in climate over long timescales, but the current warming is driven primarily by the rapid accumulation of greenhouse gases, particularly CO_2, methane (CH_4), and nitrous oxide (N_2O). The increase in atmospheric CO_2 levels, from around 280 parts per million (ppm) before the Industrial Revolution to over 400 ppm today, is not part of a natural cycle but rather the result of human actions. The burning of fossil fuels for energy, transportation, and industry has released large quantities of CO_2 into the atmosphere, while deforestation has reduced the Earth's capacity to absorb CO_2, further exacerbating the problem. This unprecedented level of human influence on the climate makes the current situation fundamentally different from past natural climate shifts, which were largely driven by geological or astronomical factors.

By understanding the natural cycles of climate and the historical impacts of past climate shifts, we can better predict and mitigate the potential effects of current climate change. Past shifts in climate provide valuable lessons in how ecosystems and societies have adapted—or failed to adapt—to changing conditions. For example, during the last Ice Age, humans adapted to changing climates by developing new technologies, shifting their ways of life, and migrating to more hospitable areas. However, the rapid pace of current climate change presents a challenge that is far more immediate and widespread, with impacts that are already being felt across the globe. Predicting the future impacts of climate change will require an understanding of both the natural variability of the climate system and the unique role that human activities play in driving and accelerating these changes.

By looking at the historical context of climate change, we can also see the importance of proactive measures. Throughout history, societies that failed to adapt to environmental changes faced significant challenges, from the collapse of ancient civilizations due to prolonged droughts to the devastation caused by rising sea levels and changing agricultural patterns. Today, we have the knowledge and technology to take action to mitigate the effects of climate change before it leads to irreversible damage. However, the window of opportunity to limit warming to 1.5°C or even 2°C above pre-industrial levels is rapidly closing, and without immediate and sustained global action, we risk crossing tipping points that could lead to catastrophic consequences.

The urgency of understanding and addressing climate change cannot be overstated. While past climate shifts offer valuable insights, the current situation is unlike anything humanity has faced before. The rapid pace of warming, coupled with the unprecedented levels of greenhouse gases in the atmosphere, demands immediate action to reduce emissions, transition to renewable energy sources, and restore ecological balance. By learning from history and applying this knowledge to modern solutions, we have the potential to mitigate the worst effects of climate change and safeguard the future of our planet.

Industrialization and Its Impact on Climate

The Industrial Revolution, beginning in the late 18th century, marked the onset of profound changes in human society and the Earth's climate, setting the stage for the modern era of climate change. Prior to this period, human societies largely depended on renewable energy sources, such as wood, water, and wind, to power their economies. These energy sources, although finite in their own right, were part of a relatively sustainable system. Agricultural production, transportation, and industry were all limited by the scale of available natural resources, which constrained their environmental impact. However, the Industrial Revolution initiated a fundamental shift in how humans sourced and used energy, and with it came an intensification of human impact on the planet's climate.

The primary catalyst of this transformation was the widespread adoption of coal as a fuel source. Coal was more energy-dense than wood and other natural fuels, enabling the industrialization of key sectors such as manufacturing, transportation, and power generation. The first major technological breakthrough was the steam engine, invented by James Watt and others in the late 18th century, which became a cornerstone of industrial power. Steam engines, initially used to pump water out of coal mines, quickly found applications in locomotives, factories, and ships, dramatically increasing production efficiency. The need for coal to fuel these machines led to large-scale mining operations, which not only intensified the extraction of fossil fuels but also catalyzed urbanization as factories were built and transportation networks expanded. This new reliance on coal marked the beginning of the fossil fuel era, one in which human energy consumption far exceeded anything seen before.

The widespread use of coal during the Industrial Revolution also had a profound impact on the Earth's climate. Burning coal releases carbon dioxide (CO_2) and other greenhouse gases into the atmosphere, marking the beginning of human-induced climate change. Prior to industrialization, CO_2 concentrations in the atmosphere were relatively stable at about 280 parts per million (ppm) for thousands of years. However, with the onset of industrial activity, particularly the combustion of coal in steam engines and factories, atmospheric CO_2 levels began to rise steadily. This increase in CO_2 trapped more heat in the Earth's atmosphere, leading to gradual warming, a process that became increasingly evident as industrialization expanded globally. While early climate models could not predict the full extent of the warming, the foundational work laid during this period set the stage for the climate science that would develop in the 20th century.

By the late 19th and early 20th centuries, the fossil fuel revolution was in full swing. The use of coal continued to rise, and new energy sources, particularly oil and natural gas, were developed, further fueling industrial growth. Oil, which had been used sparingly in lamps, began to replace coal in a variety of applications, from transportation (with the invention of the automobile and the growth of the oil industry) to the production of chemicals, plastics, and fertilizers. The introduction of natural gas as a cleaner-burning alternative to coal in power generation added another layer to this growing reliance on fossil fuels. By the mid-20th century, human societies were deeply dependent on fossil fuels for almost every aspect of their economies, from electricity generation to industrial production and transportation.

Fossil Fuels and the Rise of CO_2 Emissions

The shift to fossil fuels, particularly coal, marked the beginning of a dramatic increase in CO_2 emissions, which has continued to accelerate over the past two centuries. The early stages of industrialization, characterized by the use of coal in steam engines and factories, significantly boosted the release of CO_2 into the atmosphere. As coal became the primary energy source for industrial processes, the scale of fossil fuel consumption grew exponentially. By the late 19th century, major cities in Europe and North America were powered by coal, and the rapid expansion of factories and railroads further increased demand for this resource.

The burning of coal, oil, and natural gas releases large quantities of CO_2, a greenhouse gas that traps heat in the atmosphere. The more CO_2 that is released, the greater the potential for the Earth's atmosphere to retain heat, leading to global warming. The rise in global temperatures is a direct consequence of the increased concentration of greenhouse gases, particularly CO_2. By the early 20th century, the levels of CO_2 in the atmosphere were already beginning to increase steadily, marking the start of what would become a long-term warming trend. This was further amplified as the global population grew and industrial activity expanded.

The global temperature record from the late 19th and early 20th centuries shows that, as CO_2 concentrations began to rise, so did global temperatures. Between 1880 and 1980, global temperatures rose by approximately 0.6°C, a modest but significant increase in just one century (Hansen et al., 1981). This marked the first major warming phase linked directly to industrialization and the burning of fossil fuels. While this warming might seem small in isolation, it represents a profound shift in Earth's climate system, one that has continued into the present day. The warming trend accelerated in the second half of the 20th century, coinciding with the continued rise in emissions from the expanding global industrial economy.

In addition to CO_2, the burning of fossil fuels also releases other greenhouse gases, such as methane (CH_4) and nitrous oxide (N_2O), both of which are even more potent than CO_2 in trapping heat in the atmosphere. The rapid expansion of transportation networks in the 20th century, particularly with the rise of the automobile and the expansion of air travel, further contributed to the release of these gases.

In fact, transportation has become one of the largest sources of greenhouse gas emissions in the modern world. Industrial activities, such as cement production and chemical manufacturing, also contribute significant amounts of CO_2, further exacerbating the problem.

The persistent increase in CO_2 and other greenhouse gases has led to a steady rise in global temperatures, with profound implications for ecosystems, weather patterns, and human societies. The acceleration of warming, particularly since the mid-20th century, is a clear indication that the Earth's climate is being significantly altered by human activity. The question now is not whether the climate is changing, but how rapidly and how severely the impacts will unfold in the coming decades.

As the global temperature continues to rise, the effects of fossil fuel emissions are becoming increasingly evident. More frequent and intense heatwaves, melting polar ice, rising sea levels, and extreme weather events are just some of the consequences of this warming. The history of industrialization and its impact on CO_2 emissions demonstrates that human activities have been a major driver of climate change, and unless significant action is taken to reduce emissions and transition to renewable energy sources, the trajectory of global warming will continue to intensify.

Deforestation and Land Use Change

In addition to the combustion of fossil fuels, industrialization has also led to widespread deforestation and significant land-use changes, both of which have played a major role in accelerating climate change. Forests, which serve as vital carbon sinks, absorb carbon dioxide (CO_2) from the atmosphere and store it in their biomass and soils. This process helps mitigate the effects of climate change by removing CO_2, one of the primary greenhouse gases responsible for global warming. However, as industrialization has expanded, the need for land for agriculture, urbanization, and resource extraction has led to large-scale deforestation, particularly in tropical regions.

In tropical rainforests like the Amazon, forests are being cleared at an alarming rate for agricultural expansion, logging, and infrastructure development. When trees are cut down and burned, not only is the carbon that was stored in the trees released back into the atmosphere, but the forest's capacity to sequester additional carbon is also lost. This combination of carbon release and reduced carbon sequestration significantly amplifies the warming effects of the rising concentrations of CO_2 in the atmosphere. It is estimated that land-use changes and deforestation account for about 23% of global CO_2 emissions (IPCC, 2021), a staggering figure that underscores the critical role forests play in the global carbon cycle.

The deforestation of the Amazon rainforest, often referred to as the "lungs of the planet," is one of the most concerning examples of how human activity is exacerbating the climate crisis. The Amazon not only sequesters vast amounts of carbon, but it also plays a crucial role in regulating global weather patterns, particularly through its influence on rainfall and temperature systems. As forests continue to be cleared, the consequences are felt not only in the form of higher CO_2 emissions but also through the destabilization of local and global weather systems. The loss of forests, particularly in such vital regions, creates a vicious cycle: the release of carbon accelerates climate change, which in turn increases the likelihood of more extreme weather events and further deforestation, leading to even more warming.

Beyond CO_2, deforestation also contributes to the loss of biodiversity, which further weakens the resilience of ecosystems to the impacts of climate change. Deforestation disrupts the balance of ecosystems, threatening species that rely on forests for habitat, food, and water. This loss of biodiversity has cascading effects on food security, water resources, and the overall health of ecosystems, further compounding the challenges of addressing climate change.

The Greenhouse Effect and Its Consequences

The industrialization of human societies, coupled with widespread deforestation and land-use changes, has significantly enhanced the natural greenhouse effect, leading to global warming. The greenhouse effect is a natural process by which certain gases in Earth's atmosphere—known as greenhouse gases (GHGs)—trap heat, keeping the planet warm enough to sustain life. Without this effect, Earth would be inhospitably cold. However, human activities, particularly the burning of fossil fuels and large-scale deforestation, have intensified this natural process, leading to an unprecedented increase in global temperatures.

The combustion of fossil fuels for energy production, transportation, and industry is the primary source of CO_2 emissions. In addition to CO_2, the burning of fossil fuels also releases methane (CH_4) and nitrous oxide (N_2O), both of which are far more potent greenhouse gases than CO_2, albeit present in smaller quantities. Methane is released primarily from agriculture, particularly livestock, as well as from the extraction and transportation of fossil fuels. Nitrous oxide, meanwhile, is emitted from agricultural practices, including the use of synthetic fertilizers. These gases contribute to global warming by trapping more heat in the atmosphere, causing the planet to warm at a faster rate.

As the concentration of CO_2, CH_4, and N_2O in the atmosphere rises, the Earth's climate system becomes increasingly unstable. The higher concentrations of these gases trap more heat, leading to a rise in global temperatures, altered precipitation patterns, and more extreme weather events, such as heatwaves, hurricanes, and flooding. CO_2, in particular, is the most significant contributor to climate change. Its levels in the atmosphere have now surpassed 400 parts per million (ppm)—a threshold not seen for millions of years (Lindsey, 2021). This unprecedented concentration of CO_2 is a direct result of human activity, particularly fossil fuel combustion and deforestation, and it has triggered a rapid shift in the Earth's climate.

The accumulation of greenhouse gases, particularly CO_2, in the atmosphere is the primary driver of the climate crisis. The increase in CO_2 levels has led to a steady rise in global temperatures, with the planet already warming by approximately 1.1°C since the late 19th century. While this might seem like a small increase, it has already had profound impacts on ecosystems, human societies, and the global economy. The rise in temperatures has led to the melting of polar ice caps, rising sea levels, and more frequent and intense extreme weather events. These changes threaten to disrupt food and water supplies, displace millions of people due to sea-level rise, and overwhelm infrastructure designed for a climate that is no longer stable.

The consequences of this enhanced greenhouse effect are already being felt across the globe. Coral reefs, which act as vital carbon sinks and protect coastal communities, are experiencing widespread bleaching due to rising ocean temperatures. The Arctic, which is warming at more than twice the rate of the global average, is seeing rapid ice melt, leading to rising sea levels that threaten coastal cities worldwide. Forests, which once acted as carbon sinks, are now being cleared or dying from droughts and fires, releasing stored carbon and exacerbating the climate crisis.

As global temperatures continue to rise, the feedback loops of the greenhouse effect become even more pronounced. For instance, as ice sheets and glaciers melt, the Earth's albedo (reflectivity) decreases, meaning that less sunlight is reflected back into space, and more heat is absorbed by the oceans and land, accelerating warming. Similarly, as forests and wetlands are destroyed, their ability to absorb CO_2 is reduced, further increasing atmospheric concentrations of greenhouse gases. These feedback loops create a self-reinforcing cycle, making it increasingly difficult to slow or reverse the warming process.

The overwhelming presence of greenhouse gases in the atmosphere is driving an unprecedented acceleration of climate change. The long-term consequences of continued emissions are dire, with projections suggesting that without significant intervention, global temperatures could rise by 2°C or more by the end of the century. This level of warming would bring about widespread disruptions to ecosystems, economies, and societies, particularly in vulnerable regions. Addressing the greenhouse effect and its consequences requires urgent action to reduce emissions, transition to renewable energy, and restore ecosystems that can act as carbon sinks, such as forests and wetlands.

Role of Industrialization in Accelerating Climate Change

The industrialization of human societies has accelerated the rate of climate change in ways previously unimaginable. Before the dawn of industrialization, the Earth's climate followed natural cycles of warming and cooling, primarily driven by astronomical factors such as changes in Earth's orbit, axial tilt, and solar radiation. These processes, known as the Milankovitch cycles, operated on timescales of thousands to tens of thousands of years, allowing ecosystems to adapt to gradual shifts in temperature. However, the advent of the Industrial Revolution in the late 18th century ushered in a new era in which human activity became the dominant force influencing the climate system, disrupting the natural balance.

The rapid industrial expansion that began with the widespread use of coal as a primary energy source marked the beginning of an unprecedented release of greenhouse gases into the atmosphere. CO_2, methane (CH_4), and nitrous oxide (N_2O)—all potent greenhouse gases—began to accumulate at rates far beyond what the planet had experienced in its natural cycles. The burning of coal in steam engines, followed by the use of oil and natural gas in power generation and transportation, led to an exponential increase in atmospheric CO_2 levels. Industrial processes, such as cement production, steel manufacturing, and chemical production, further contributed to the accumulation of these gases. By the 20th century, industrial emissions had become the primary driver of the climate crisis, setting in motion a cycle of disruption that continues to this day.

Alongside the combustion of fossil fuels, industrialization has also resulted in massive land-use changes, particularly through deforestation. Forests, which act as vital carbon sinks, are being cleared to make way for agriculture, urbanization, and infrastructure development. This not only reduces the Earth's capacity to absorb CO_2 but also releases the carbon stored in trees back into the atmosphere when they are cut down and burned. The loss of forests, particularly tropical rainforests like the Amazon, has exacerbated the effects of greenhouse gas emissions, creating a feedback loop that accelerates global warming. The destruction of these ecosystems further destabilizes the climate, leading to the loss of biodiversity, the depletion of natural resources, and the disruption of essential ecological services such as water filtration, pollination, and climate regulation.

The consequences of industrialization on the climate are already evident. Global temperatures have risen by approximately 1.1°C since the late 19th century, with projections indicating that the Earth could warm by an additional 2°C or more by the end of the century if emissions continue at their current rate. This warming is having a cascading effect on the planet's ecosystems and weather systems. Glaciers and polar ice caps are melting, contributing to rising sea levels that threaten coastal cities and communities. Extreme weather events, such as hurricanes, heatwaves, droughts, and floods, are becoming more frequent and severe, impacting millions of people worldwide. Ecosystems are being disrupted, and species are being driven to extinction at an alarming rate. The industrialization of human societies has created a climate system that is increasingly unstable and prone to catastrophic events, making the need for action more urgent than ever.

The legacy of industrialization and its profound impact on climate change is not merely a historical matter but a current and future challenge. The industrial revolution set in motion a cycle of climate disruption that continues to intensify, with rising temperatures, melting ice, and shifting weather patterns. As we look to the past for insights, it is clear that the choices we make today will determine the future of our planet. While efforts to mitigate these impacts are underway, the question remains: can humanity reduce emissions quickly enough to avoid the most catastrophic consequences of climate change? The answer lies in our ability to learn from history, transition to renewable energy, and implement strategies to reduce emissions and restore ecological balance. The time to act is now—by embracing innovation, taking decisive action, and restoring the balance that has been disrupted, we can mitigate the worst effects of climate change and create a sustainable future for generations to come. Science is clear, the stakes are high, and the responsibility to act rests with us all.

The urgency of addressing climate change is not a matter of speculation; it is a reality that we are already facing. If we fail to act, the consequences will be felt across the globe, through more extreme weather, rising sea levels, loss of biodiversity, and irreversible damage to ecosystems and human societies. The legacy of industrialization has set the stage for the current crisis, but it also provides us with the tools and knowledge to turn the tide. It is up to us to embrace innovation, accelerate the transition to renewable energy, and take meaningful steps to reduce emissions. The future of our planet depends on the choices we make today, and the time to act is now.

Climate science continues to evolve, integrating disciplines like oceanography, atmospheric chemistry, and satellite remote sensing. But its foundation remains grounded in the work of early pioneers who first suspected that the atmosphere was more than just empty air.

Understanding the development of climate science reveals not only the rigor of its methods but also the urgency of its findings. This chapter is not just about the past; it is about the responsibility that comes with knowledge. What we now know, we cannot unknow. The next chapters must focus not only on continued observation but on action.

Chapter 3

The Human Factor

Understanding climate change goes beyond the realm of scientific data and models—it also requires examining how human psychology, behavior, and public perception shape our responses to this global crisis. Over time, climate change has been perceived in different ways, influenced by cultural, political, and economic contexts. From early scientific inquiries to the present-day recognition of its urgency, the way society frames climate change has evolved significantly. In this chapter, we will explore how climate change awareness and action have been shaped by cognitive biases, political ideologies, and emotional responses. Additionally, we will examine how public perceptions have shifted over time, particularly with the rise of environmental movements and youth-led activism. By understanding the psychological and behavioral factors at play, we can better understand the barriers to addressing climate change and the strategies needed to drive effective action.

Examining the Historical Context of Climate Change Perception

Climate change has long been a topic of interest, but its public perception and scientific understanding have evolved significantly over time. Early conceptions of climate change were largely based on observations of weather patterns and seasonal variations, without the realization that these patterns were linked to long-term global shifts. The idea that human activities could influence the climate began to take shape in the 19th century, with the advent of scientific inquiry into atmospheric processes. In 1824, Joseph Fourier, a French mathematician, proposed the revolutionary idea that the Earth's atmosphere acted like a blanket, trapping heat and maintaining a temperature suitable for life. This early concept of the "greenhouse effect" laid the foundation for future climate science (Fourier, 1824). However, it was not until the late 19th and early 20th centuries that scientists began to connect human activities, particularly the burning of fossil fuels, to long-term changes in the climate.

In the 20th century, the recognition of climate change's potential impacts began to grow as industrialization spread across the globe, and human activities increasingly contributed to higher concentrations of greenhouse gases. During this period, early environmental concerns about pollution and deforestation began to intersect with climate change discourse, particularly in the mid-20th century, when studies started to link rising carbon dioxide (CO_2) levels with warming global temperatures. The 1970s saw the birth of modern environmentalism, spurred on by seminal events such as the first Earth Day in 1970, which brought environmental issues, including air pollution and the emerging climate crisis, into the public consciousness (McKibben, 1989).

As the awareness of climate change grew, so did the urgency with which scientists and activists called for action. By the late 20th century, scientific reports from the Intergovernmental Panel on Climate Change (IPCC) solidified the view that human activities were contributing significantly to climate change. However, the public's understanding of the issue was slow to catch up. The media played a key role in shaping the narrative, often framing climate change as a distant problem, leaving the public with a limited sense of urgency.

Historical Events and Natural Shifts

Throughout history, the framing of climate change has been influenced not only by scientific discoveries but also by major historical events and natural climate shifts. During periods of cooling, such as the Little Ice Age (approximately 1300–1850), the Earth's natural climate variability was much more pronounced. These cold periods, marked by crop failures and harsh winters, created a heightened awareness of the impacts of climate on human livelihoods. The Little Ice Age, while not directly caused by human activity, shaped the early understanding of how significant climate shifts could affect societies, contributing to public and scientific perceptions of the climate as a variable force.

The framing of climate change also owes much to the rise of environmental movements in the 20th century. These movements, spurred by growing concerns about pollution and deforestation, gradually incorporated climate change into their agendas. Notable milestones in this evolution include the establishment of the Kyoto Protocol in 1997, which set binding emission reduction targets for industrialized nations, and the Paris Agreement in 2015, which sought to limit global warming to well below 2°C above pre-industrial levels. These global agreements highlighted the growing recognition of climate change as a global, human-induced crisis that required international cooperation and immediate action.

Cognitive Biases and Human Behavior in Climate Action

While scientific evidence and data on climate change have accumulated over decades, psychological barriers continue to hinder meaningful action. Cognitive biases, which are patterns of thought that deviate from rational judgment, often prevent individuals and societies from fully acknowledging the severity of the climate crisis. Denial is one of the most pervasive cognitive biases when it comes to climate change. For many, the reality of climate change is difficult to accept, especially when its most severe consequences seem distant or abstract. Some deny the scientific consensus on human-induced climate change altogether, while others downplay the urgency of the problem.

Optimism bias, another common cognitive bias, causes individuals to underestimate the risks associated with climate change because they assume that future events will unfold more favorably than the evidence suggests. This bias leads people to believe that technology or future innovations will mitigate the problem, without immediate action being necessary. Temporal discounting, the tendency to favor short-term rewards over long-term benefits, exacerbates this bias by making the immediate benefits of carbon-intensive activities, such as using fossil fuels, more appealing than the long-term consequences of inaction (Gifford, 2011).

Another psychological barrier is the fragmentation of climate change as a political issue. Climate change is often framed through ideological lenses, with political identity influencing individuals' perceptions of the problem. In many countries, particularly the United States, climate change has been politicized, with significant portions of the population perceiving it as a partisan issue rather than a global one that requires collective action. This polarization is a major obstacle to enacting the broad, systemic change needed to address the crisis.

Role of Emotions, Short-Term Thinking, and Ideologies

Emotions also play a central role in how individuals perceive climate change and respond to it. Fear, anxiety, and guilt can often paralyze action rather than motivate it. Some individuals, faced with the overwhelming evidence of climate change, may experience eco-anxiety—an emotional response to the perceived loss of the environment and its potential effects on future generations. This anxiety can lead to feelings of helplessness or even apathy, as people may feel powerless to make a meaningful difference. Conversely, emotions like hope and empowerment, particularly when people feel they can make a difference, can foster positive action (Clayton, 2020).

The combination of short-term thinking and ideological divides creates a significant challenge for addressing climate change. People are often more concerned with immediate issues like economic growth, job security, and political stability, which makes it difficult to focus on long-term environmental threats. Overcoming these barriers requires not only changing attitudes but also creating incentives that make climate action more appealing in the short term, such as clean energy solutions that provide economic opportunities.

Shifting Attitudes Toward Climate Change

Public perceptions of climate change have undergone a dramatic transformation over the past few decades. Initially, climate change was viewed as a scientific issue discussed primarily in academic and policy circles, with little direct engagement from the general public. However, starting in the 1970s, and gaining momentum throughout the 1990s and 2000s, climate change became an issue of global concern, fueled by growing scientific evidence, media coverage, and grassroots activism. The first Earth Day in 1970, for instance, marked a significant turning point, as it helped solidify environmental concerns, including climate change, on the global agenda (McKibben, 1989).

By the 1990s, the scientific consensus on climate change had become clear, and global negotiations began, culminating in the Kyoto Protocol, which, although criticized for its lack of enforcement mechanisms, established climate change as an urgent global issue. Since then, the issue has evolved from a niche environmental concern to a central issue in international politics, business, and society. As the effects of climate change became more visible through rising sea levels, extreme weather events, and heatwaves, the demand for action grew, and climate change became a defining issue in global politics and economics.

Addressing and Investing in Climate Research

Despite the growing awareness and evidence surrounding climate change, political challenges persist in addressing and investing in climate research and solutions. One of the key hurdles is the resistance from political factions, particularly in countries where vested interests—such as fossil fuel industries—are tied to the status quo. In many cases, political leaders are reluctant to adopt policies that could disrupt economic growth or challenge powerful corporate interests, particularly in sectors that contribute to greenhouse gas emissions. This political resistance has hindered the implementation of comprehensive climate policies and slowed the pace of critical investments in climate research, renewable energy, and sustainable infrastructure.

Climate change, as a political issue, is often clouded by ideology. In some regions, the framing of climate action as a political agenda has led to polarized views, with one side arguing for climate action as essential for the survival of future generations, and the other side seeing it as an unnecessary economic burden. This ideological divide has slowed the political will to invest in and fund climate research. Additionally, climate change skepticism, often propagated by media and interest groups, has made it harder to secure the widespread support needed to tackle the issue effectively.

However, public movements and grassroots activism are changing this landscape. Movements like Fridays for Future and Extinction Rebellion are pushing political leaders to take stronger, more immediate action. The youth-led Fridays for Future movement, which began with Greta Thunberg's school strikes, has gained global traction, influencing policy discussions and demanding investments in climate research, clean energy, and global action on climate change. This shift is helping to break down the political barriers that have historically hindered progress and pushing the issue of climate change to the forefront of global discourse.

The Rise of Movements and the Role of Media

In the 21st century, the rise of youth-led movements like Fridays for Future and grassroots campaigns such as Extinction Rebellion have been instrumental in shifting public perceptions and pushing governments and industries to act. These movements emphasize the urgency of climate action and challenge the status quo by demanding accountability from political leaders and corporations. By framing climate change as an issue of justice—intergenerational justice, environmental justice, and social justice—these movements have brought climate change into the spotlight, especially among younger generations who feel the weight of future consequences. Through strikes, protests, and campaigns, these activists have catalyzed a broader public awakening to the issue, illustrating the power of grassroots activism in driving policy change.

The media has played a pivotal role in this shift, with growing coverage of climate change in news outlets, documentaries, and social media platforms. Media coverage, particularly after extreme weather events, has helped connect the dots between these events and climate change, making the issue more tangible for the public. Public figures and celebrities, such as Greta Thunberg and Leonardo DiCaprio, have also used their platforms to advocate for urgent climate action, further raising awareness. These shifts in perception are critical for driving the systemic changes needed to tackle the climate crisis.

As we confront the escalating climate crisis, it is clear that addressing the psychological barriers, shifting public perceptions, and overcoming political challenges are as crucial as the scientific and technological solutions. Our understanding of climate change has evolved, but the path to meaningful action requires more than just awareness—it demands a collective, urgent response driven by a fundamental shift in how we think, behave, and act as individuals and societies. To fully understand the gravity of climate change, we must first examine its history, tracing the evolution of climate science, the role of industrialization, and the pivotal moments that have shaped public perception and action. In the following chapters, we will delve deeper into these historical contexts, exploring how past climate shifts have influenced current perceptions and the urgent need for action. The time to break through cognitive biases, confront ideological divides, and embrace global cooperation is now. Only by aligning our actions with the overwhelming evidence and urgency of the crisis can we hope to secure a sustainable future for all. The power to change lies within us, and the choices we make today will define the planet for generations to come.

Chapter 4

When Europe Froze

The Alpine Glaciation, a defining chapter in Earth's climatic history, profoundly shaped the European landscape and the ecosystems that inhabited it. Originating millions of years ago, this glaciation was part of the larger Quaternary glaciations that left an indelible mark on the continent. This chapter delves into the origins and timeline of the Alpine Glaciation, its impacts on landscapes and ecosystems, the geological evidence of glacial advances and retreats, and the remarkable adaptations of early humans and animals to these icy conditions.

Origins and Timeline of the Alpine Glaciation

The Alpine Glaciation, part of the Pleistocene Epoch (2.58 million to 11,700 years ago), was a defining period of climatic and geological transformation, characterized by cycles of glacial advance and retreat. These cycles were primarily driven by Milankovitch cycles—variations in Earth's orbit, axial tilt, and precession—that alter the distribution and intensity of solar radiation reaching the planet's surface (Hays, Imbrie, & Shackleton, 1976). These orbital variations initiated climatic shifts that resulted in extended periods of ice coverage across the Alpine region. Glaciation occurred not as a single, continuous event but as a series of episodic advances and retreats, with ice sheets expanding during colder phases and retreating during interglacial warm periods.

The first significant Alpine glaciation began during the Early Pleistocene, roughly 2.4 million years ago. However, it was in the Middle to Late Pleistocene, between approximately 800,000 and 11,700 years ago, that the region experienced its most pronounced glacial activity. These later glaciations were associated with a combination of intensified cooling and an accumulation of snow and ice in the high-altitude regions of the Alps. With each glacial cycle, massive ice sheets extended beyond the mountainous areas, reaching into the surrounding valleys and lowlands, altering landscapes and ecosystems on a vast scale.

Among the multiple glaciations that occurred, the Würm glaciation stands out as the last and most extensively studied phase of the Alpine Glaciation. This glacial phase began approximately 115,000 years ago and coincided with a global cooling event that heralded the start of the last Ice Age. The Würm glaciation reached its peak around 20,000 years ago, during the Last Glacial Maximum (LGM), a period of maximum global ice coverage. At its height, glaciers originating from the Alpine highlands advanced into the foothills and surrounding plains, profoundly reshaping the European continent. The ice masses carved deep valleys, transported enormous quantities of sediment, and left behind prominent landforms such as moraines and fjords (Ehlers & Gibbard, 2007).

The end of the Würm glaciation, approximately 10,000 years ago, marked a significant turning point in Earth's climatic history. As the glaciers began to retreat, the Pleistocene transitioned into the Holocene Epoch, a period characterized by a warmer and more stable climate. This retreat was not instantaneous but occurred gradually over several thousand years, with intermittent phases of re-advance and stasis depending on localized climatic fluctuations. The melting of these massive ice sheets dramatically altered the Alpine and European landscapes, creating lakes, braided river systems, and fertile soils that would later support human agricultural development.

The retreat of the glaciers also had profound ecological implications. The warming climate allowed forests and grasslands to recolonize previously glaciated areas, enabling the reestablishment of ecosystems that had been displaced to southern refugia during the glacial maximum. This new environmental stability provided the conditions necessary for the rise of early human civilizations, as it supported a reliable food supply and the development of permanent settlements.

The Würm glaciation's legacy is still visible today in the Alpine region, where the striking glacial landscapes bear testament to the immense power of ice in shaping the Earth's surface. These geological features, combined with the climatic transitions that followed, offer valuable insights into the complex interplay of natural forces that govern our planet's climate system. This understanding not only sheds light on the past but also informs our ability to predict and respond to future climate changes.

Impacts on the European Landscape and Ecosystems

The Alpine Glaciation profoundly shaped the European landscape, leaving behind a legacy of dramatic physical features and reshaped ecosystems. The immense pressure and movement of glaciers acted as powerful agents of erosion, carving out U-shaped valleys, sharp peaks, and deep fjords. These glacial landforms remain some of the most iconic features of the region, with landmarks such as the Matterhorn and Mont Blanc standing as enduring symbols of the Alpine range's geological history. The relentless grinding of ice against rock sculpted sheer cliffs and steep ridges, while glacial plucking and abrasion smoothed valley floors, creating the classic U-shaped profiles now associated with glaciated mountain regions.

The impact of glacial erosion extended beyond the highlands into the lowlands, where glaciers deposited vast quantities of sediment, reshaping the surrounding terrain. This process created fertile plains and sedimentary basins, some of which are now sites of major human settlement and agricultural activity. Additionally, as glaciers retreated, they left behind depressions that filled with meltwater, forming a series of lakes, including Lake Geneva, Lake Como, and Lake Maggiore. These glacial lakes not only continue to define the region's natural beauty but also play vital roles in local water cycles, ecosystems, and economies, supporting tourism, agriculture, and hydroelectric power generation (Benn & Evans, 2010).

The Alpine Glaciation also had profound effects on ecosystems, forcing flora and fauna to adapt to the shifting and often harsh conditions. As ice sheets expanded, they displaced forests and grasslands, pushing vegetation zones southward into refugia—isolated pockets where species could survive under glacial conditions. These refugia, located in sheltered valleys or along southern slopes, served as sanctuaries for biodiversity. During interglacial periods, as the ice retreated and the climate warmed, these areas became sources of recolonization, with species migrating back into previously glaciated regions (Tzedakis et al., 2013).

The repeated advance and retreat of glaciers not only displaced ecosystems but also fragmented habitats, driving evolutionary processes that led to the emergence of unique, cold-adapted species. The Alpine ibex, with its remarkable climbing ability and specialized diet, is a prime example of a species that evolved to thrive in the rugged, glaciated landscapes. Similarly, the mountain hare developed a seasonal coat that changes color for camouflage in snow-covered or rocky environments, enhancing its survival in the cold alpine conditions.

Glacial cycles also influenced the genetic diversity and distribution of plant species. Cold-tolerant species, such as the edelweiss and glacier buttercup, adapted to thrive in the thin soils and extreme climates of high-altitude regions, becoming emblematic of Alpine flora. These plants not only survived but also contributed to soil stabilization and nutrient cycling, aiding the recovery of ecosystems during interglacial periods.

Beyond individual species, the glacial legacy shaped entire ecosystems. Proglacial environments—areas at the edge of retreating glaciers—became hotspots of dynamic ecological activity. These areas, characterized by newly exposed rock and sediment, served as laboratories of ecological succession, where pioneering species like lichens and mosses established the foundation for more complex plant and animal communities. As ecosystems developed, they provided critical habitats for a wide range of organisms, contributing to the overall resilience and biodiversity of the Alpine region.

The enduring influence of the Alpine Glaciation is visible not only in the landscapes it sculpted but also in the ecological patterns and processes it set into motion. These changes continue to shape human and ecological interactions in the region, offering invaluable insights into the interplay between climatic forces and life on Earth. The study of these impacts underscores the importance of understanding glacial history, not just for its geological significance but also for its role in driving evolutionary and ecological transformations that persist to this day.

Alpine Glacial Advances and Retreats

The evidence of Alpine glaciation is imprinted across the European landscape, offering a wealth of information about glacial dynamics and their far-reaching impacts. One of the most visible markers of glaciation is the presence of moraines—ridges of rock, gravel, and sediment deposited at the edges of glaciers. These features serve as clear indicators of the extent of glacial advances, with terminal moraines marking the farthest reach of ice sheets during specific glacial periods. For example, the terminal moraines in the Rhine Valley and the Po Basin provide tangible evidence of the maximum extent of the Würm glaciation, helping geologists map the impressive scale of Alpine ice coverage (Ehlers et al., 2011).

Another compelling feature of glacial activity is the presence of striations—linear scratches and grooves etched into bedrock by the abrasive movement of glaciers laden with rocks and debris. These striations not only reveal the immense power of glacial forces but also provide information about the direction and flow patterns of the ice. By studying these markings, researchers can reconstruct the pathways glaciers took as they advanced and retreated across the landscape, adding depth to our understanding of glacial mechanics.

Sedimentary deposits left behind by glaciers further enrich the geological record of Alpine glaciation. Glacial till, a heterogeneous mixture of clay, silt, sand, and boulders deposited directly by glaciers, blankets many areas that were once ice-covered. This till often forms drumlins—elongated hills that provide additional clues about glacial movement. Beyond till, outwash plains, composed of sediments carried by meltwater, stretch across lowland areas, demonstrating the dynamic interplay between glaciers and hydrology. These plains, formed by braided river systems created during interglacial periods, showcase the transformative impact of meltwater as it redistributed sediments and reshaped the terrain.

Proglacial lakes, another hallmark of glacial retreat, were formed in depressions left by melting ice or blocked by moraines. These lakes, such as Lake Geneva and Lake Maggiore, are not only lasting features of the Alpine region but also serve as valuable archives of climatic and environmental changes. Sediments accumulating in these lakes over thousands of years preserve records of vegetation, erosion, and water chemistry, offering insights into past climatic conditions.

Speleothems—stalactites and stalagmites found in Alpine caves—provide an indirect yet highly informative record of glacial and interglacial periods. These formations grow as water carrying dissolved minerals drips through cave ceilings, creating layers that reflect variations in temperature and precipitation. Isotopic analyses of speleothems, such as ratios of oxygen isotopes, allow scientists to infer past climate conditions, including glacial extent and meltwater contributions (Fairchild & Baker, 2012). This data complements other geological evidence, creating a more comprehensive picture of Alpine glaciation.

Collectively, these geological records offer a detailed narrative of the Alpine glaciation, documenting not only the physical processes involved but also the timing and extent of glacial advances and retreats. Beyond their historical significance, these records provide critical data for reconstructing past climates. By analyzing the patterns and impacts of previous glacial cycles, scientists can identify trends and feedback mechanisms that inform our understanding of current and future climate dynamics in glacial regions.

The significance of this evidence extends beyond the scientific community. As modern climate change accelerates glacier retreat, the study of these ancient features offers vital insights into the potential consequences of ice loss, including impacts on hydrology, ecosystems, and human populations. This geological legacy serves as both a warning and a guide, reminding us of the profound influence of glacial activity on the Earth's surface and the importance of preserving these records for future generations.

Early Human and Animal Adaptations

The harsh and unforgiving conditions of the Alpine Glaciation presented significant survival challenges for early humans and animals, spurring remarkable adaptations that highlight the resilience and ingenuity of life during the Pleistocene Epoch. For humans, survival meant not only enduring extreme cold but also navigating the shifting landscapes and scarce resources of a glacial environment. Neanderthals, who occupied Europe for hundreds of thousands of years, exemplified this adaptability. Their robust physiques—short, stocky builds with large nasal cavities—helped retain body heat and efficiently warm inhaled air, critical for surviving the frigid climate (Zilhão, 2001). Neanderthals' technological innovations included advanced tool-making, such as the creation of Mousterian tools, which were crafted from stone and used for hunting and processing large mammals like woolly mammoths and reindeer. These animals were integral to Neanderthal survival, providing food, hides for clothing, and bones for tools.

The arrival of anatomically modern humans (Homo sapiens) in Europe around 40,000 years ago brought new and sophisticated strategies for coping with the challenges of the Alpine Glaciation. Unlike their Neanderthal counterparts, Homo sapiens developed tailored clothing made from animal hides and furs, stitched together using bone needles. These garments offered superior insulation, allowing them to venture further into icy environments. Shelters became more advanced, ranging from structures built from mammoth bones to semi-subterranean dwellings insulated with earth and vegetation. These innovations enabled Homo sapiens to expand their range and exploit diverse environments.

Artistic expression, as evidenced by cave paintings such as those in Chauvet Cave in modern-day France, offers a window into the lives of these early humans. The detailed depictions of animals like mammoths, horses, and aurochs not only demonstrate a deep connection to the glacial landscape but also suggest symbolic or spiritual practices tied to survival and adaptation (Clottes, 2008). Such cultural innovations may have strengthened group cohesion, critical for enduring harsh conditions and competing for limited resources.

Animals, too, displayed extraordinary adaptations to the glacial environment. Large mammals like the woolly rhinoceros and steppe bison developed thick fur and substantial fat reserves to insulate against freezing temperatures. Their physiological traits, such as low surface-area-to-volume ratios, minimized heat loss, while their foraging behaviors allowed them to subsist on sparse vegetation buried under snow. Smaller mammals, such as lemmings and voles, utilized burrowing behaviors to escape the cold and avoid predators, creating insulated environments that protected them during the harshest winters.

Bird species displayed adaptive migration strategies, traveling vast distances to find suitable habitats as glacial ice sheets expanded and contracted. This periodic movement between breeding and overwintering grounds allowed species to exploit seasonal resources, ensuring their survival during the glaciation. The adaptive behaviors of these animals highlight the interconnectedness of life during this era, with ecosystems reorganizing in response to the advancing and retreating ice.

The glacial period also saw the development of symbiotic relationships between humans and animals, which became pivotal for survival. One of the most significant examples is the domestication of dogs, thought to have begun during the Pleistocene. These early canines likely assisted Homo sapiens in hunting by tracking prey or driving animals into traps, while also providing companionship and contributing to group security (Perri, 2019). This partnership not only improved hunting efficiency but also fostered a bond that would evolve into one of humanity's most enduring relationships with another species.

The Alpine Glaciation was a transformative epoch that tested the limits of life, fostering remarkable adaptations and innovations that reshaped the European landscape, ecosystems, and the course of human evolution. From the robust physiques of Neanderthals and the cultural ingenuity of Homo sapiens to the evolutionary traits of glacial fauna, life during this period demonstrated unparalleled resilience in the face of extreme environmental pressures. These challenges acted as a crucible for survival strategies that not only ensured persistence but also laid the foundation for interconnected ecosystems and human advancement. The enduring marks of the Alpine Glaciation, visible in the physical landscape and the evolutionary legacy it left behind, remind us of nature's immense power to challenge and transform. This period stands as a testament to the adaptability of life, offering timeless lessons for navigating today's changing world with resilience and innovation.

Chapter 5

Britain in Ice

Glaciation of Great Britain during the Pleistocene Epoch was a transformative period that left an enduring impact on the region's geography, ecosystems, and early human inhabitants. Ice sheets periodically advanced and retreated across the British Isles, reshaping its landscapes and establishing links to broader glacial activity across Europe. This chapter explores the extent and timeline of glaciation in Great Britain, the profound geographical changes brought about by the ice, the post-glacial ecological and human adaptations, and the connections between British and European glaciations.

Glaciation of Great Britain: Extent and Timeline

The Pleistocene Epoch (2.58 million to 11,700 years ago) was a period of profound climatic variability, marked by repeated cycles of glacial and interglacial periods. These cycles, driven by Milankovitch cycles—predictable variations in Earth's orbital eccentricity, axial tilt, and precession—dictated the amount and distribution of solar radiation reaching the planet, leading to alternating phases of global cooling and warming (Hays, Imbrie, & Shackleton, 1976). During cooler periods, vast ice sheets expanded from polar regions and high-altitude zones, including over the British Isles, dramatically altering the region's geography and ecosystems.

The first major glaciation of the British Isles occurred during the Anglian Stage, approximately 450,000 years ago, when ice sheets extended as far south as the Thames Valley, covering much of modern-day England. This glaciation represented one of the most extensive advances of ice across the region, leaving behind prominent geological features such as tills, moraines, and meltwater channels that are still visible today. Following the Anglian, later glacial stages such as the Wolstonian (150,000–130,000 years ago) and Devensian (115,000–11,700 years ago) marked subsequent cycles of ice advance and retreat, each influenced by shifts in global temperature and precipitation patterns (Clark et al., 2012).

The Devensian glaciation, the most recent and extensively studied phase, coincided with the Last Glacial Maximum (LGM) approximately 20,000 years ago. During this period, massive ice sheets originating in Scandinavia and Scotland merged to blanket the northern and central parts of the British Isles. These ice sheets reached into Wales, the Midlands, and parts of Northern Ireland, leaving behind a landscape carved by glaciers and dotted with features like drumlins, eskers, and glacial valleys. Southern England, though not directly ice-covered, experienced harsh periglacial conditions. Intense cold caused ground freezing and thawing cycles, creating distinctive landforms such as patterned ground, ice wedges, and solifluction terraces. These frost-driven processes significantly altered the topography and soil structure, impacting both the environment and early human activity in the region.

The retreat of the ice sheets from Great Britain began around 19,000 years ago, as rising global temperatures signaled the end of the Last Glacial Maximum. This deglaciation process unfolded gradually, with episodic phases of re-advance during colder intervals. Meltwater from retreating glaciers created a network of rivers and proglacial lakes, reshaping the lowlands and contributing to the development of fertile soils in areas like East Anglia. By approximately 11,000 years ago, as the climate transitioned into the Holocene Epoch, the ice had largely disappeared, giving way to a temperate climate.

This shift marked the beginning of a new era of ecological and cultural transformation. The warming climate allowed vegetation to recolonize the landscape, starting with hardy pioneer species like mosses and lichens, followed by grasses, shrubs, and eventually deciduous forests. These changes created habitats for a diverse range of animal species and set the stage for the expansion of human populations. Mesolithic hunter-gatherers adapted to this post-glacial environment, utilizing newly available resources and establishing settlements in areas that had previously been uninhabitable.

The Pleistocene glaciations left an indelible mark on the British Isles, shaping not only its physical geography but also its ecological and cultural history. The cycles of ice advance and retreat illustrate the dynamic interplay between climate systems and terrestrial environments, providing valuable insights into past climate variability and its long-term impacts on landscapes and societies. These lessons remain relevant today as we seek to understand and navigate the complexities of a rapidly changing climate.

How Ice Shaped the British Isles' Geography

The glaciation of Great Britain profoundly transformed its geography, leaving behind a landscape characterized by striking physical features that continue to define the region today. The relentless movement of ice sheets during the Pleistocene carved out U-shaped valleys, deepened existing river channels, and scoured bedrock, creating a range of glacial landforms that bear testament to the power of ice. In Scotland, iconic features such as Glen Coe, a steep-sided valley, and the Great Glen, a linear geological fault system, were sculpted through glacial erosion. These landscapes, now central to the region's identity, reveal the pathways of massive ice flows that once dominated the terrain. Similarly, the Lake District's rugged topography, with its dramatic peaks and ribbon lakes like Windermere and Ullswater, owes much of its current form to glaciation. The glaciers that occupied this region eroded valleys and deposited debris, creating a picturesque and geologically diverse area that remains a focal point for both scientific study and tourism.

Moraines, ridges of sediment deposited at glacier margins, are prominent throughout Great Britain and provide valuable clues to the extent of past glacial activity. Terminal moraines, such as those found in the Vale of Eden, indicate the maximum reach of glaciers during specific stages of glaciation. These features serve as natural markers, offering insight into the dynamics of ice advance and retreat. Beyond moraines, drumlins—streamlined, elongated hills formed under glacial ice—are scattered across the British landscape, particularly in regions like County Down, Northern Ireland, and Lancashire, England. These features not only showcase the direction and flow of ice sheets but also provide evidence of the immense pressures exerted by moving glaciers (Benn & Evans, 2010).

In addition to erosion, glacial deposition played a pivotal role in shaping the geography of the British Isles. As glaciers retreated, they left behind thick layers of till, or glacial drift—a mix of clay, sand, gravel, and boulders deposited directly by the ice. This sediment contributed significantly to the fertility of soils in regions such as eastern England and southern Scotland, enhancing their suitability for agriculture. Meltwater from retreating glaciers further reshaped the landscape by creating outwash plains, where sediment was deposited by flowing water, and braided river systems, which evolved as meltwater streams intertwined and shifted across the lowlands. These hydrological processes transformed flat plains into complex networks of rivers and wetlands that remain prominent features of the British landscape.

The effects of glaciation also extended to coastal regions, where processes such as isostatic rebound—caused by the gradual rise of land after the removal of the weight of glaciers—altered sea levels and reshaped coastlines. Raised beaches, such as those along the west coast of Scotland, and dramatic sea cliffs, such as the Seven Sisters in southern England, are direct results of these glacial and post-glacial processes. Isostatic rebound continues to influence the region, with parts of Scotland rising by several millimeters annually while southern England experiences relative subsidence, reflecting the long-term effects of glaciation.

Perhaps one of the most dramatic impacts of glaciation on the British Isles was the formation of the English Channel. During glacial periods, when sea levels dropped significantly due to the vast quantities of water locked in ice sheets, a land bridge known as Doggerland connected Britain to continental Europe. This connection allowed the migration of animals and early humans, facilitating the exchange of species and cultural practices. Archaeological evidence from Doggerland indicates it was a rich habitat for human and animal life, featuring rivers, forests, and abundant game. However, as the ice melted and sea levels rose during the Holocene, Doggerland was gradually submerged, transforming Britain into an island and isolating its ecosystems and human populations (Gaffney et al., 2009).

The transformation of Great Britain's geography by glaciation is not just a story of past events but a continuing legacy. These glacial landforms and processes provide a rich record of Earth's climatic history and offer valuable insights into the forces that have shaped the natural world. From the dramatic valleys and lakes to the fertile soils and evolving coastlines, the imprint of the ice age remains a defining feature of the British Isles, influencing its ecology, economy, and cultural heritage.

Post-Glacial Effects on Ecosystems and Early Inhabitants

The retreat of ice sheets and the warming climate during the Holocene (beginning approximately 11,700 years ago) marked a period of profound ecological and cultural transformation in the British Isles. As temperatures rose and glaciers receded, the landscape underwent a rapid ecological succession, with vegetation recolonizing the exposed and nutrient-rich soils left behind by the ice. Pioneer species like mosses and lichens were the first to establish themselves, stabilizing the soil and creating conditions for other plants to take root. These were soon followed by grasses and shrubs, which further enriched the environment. Eventually, dense forests dominated by birch, oak, and hazel emerged, covering much of the British Isles. This reforestation created diverse habitats that supported a wide range of animal species, including red deer, wild boar, and beavers, all of which flourished in the new temperate environment. The expanding forests not only supported biodiversity but also played a critical role in shaping the human-environment interactions of the period.

For early human inhabitants, the transition from the glacial conditions of the Pleistocene to the warmer and more stable climate of the Holocene brought both opportunities and challenges. During the colder glacial periods, humans had predominantly inhabited the periglacial zones of southern England, where they relied on hunting large mammals such as reindeer, mammoths, and woolly rhinoceroses. These animals provided not only food but also materials for tools, clothing, and shelter. However, as the ice retreated and forests replaced open tundra and grasslands, these large mammals either migrated northward or became extinct, necessitating a shift in human subsistence strategies. Mesolithic communities adapted to the changing environment by diversifying their diets and resource-gathering methods. They began hunting smaller game such as red deer and wild boar, fishing in rivers and coastal waters, and gathering edible plants, nuts, and berries (Coles, 2002). This shift in subsistence strategies is indicative of the resilience and adaptability of early human populations in response to environmental changes.

Rising sea levels during the post-glacial period significantly impacted human settlement patterns. As the ice sheets melted, vast amounts of water were released into the oceans, causing sea levels to rise by approximately 120 meters globally. This process submerged large areas of low-lying land, including Doggerland, a once-thriving land bridge that connected Britain to mainland Europe. Doggerland had been a rich and fertile area, featuring rivers, forests, and abundant wildlife, and it served as a critical corridor for human and animal migration. Its gradual inundation, which began around 10,000 years ago and continued over millennia, forced communities to relocate to higher ground, reshaping the cultural geography of the British Isles. The isolation of Britain as an island around 8,000 years ago profoundly influenced its ecosystems and human societies, fostering distinct cultural and ecological developments.

Archaeological sites such as Star Carr in Yorkshire provide valuable insights into the lifestyles of these early inhabitants during the Mesolithic period. Star Carr, located near a now-vanished lake, reveals a rich material culture that included wooden platforms, flint tools, and decorative artifacts such as antler headdresses, which may have been used in ritual practices. The site also provides evidence of early woodworking techniques and the construction of dwellings, highlighting the ingenuity of these communities in adapting to their environment. These findings suggest that early inhabitants of the British Isles were not only resourceful but also deeply connected to the natural world, utilizing the abundance of post-glacial resources to establish thriving settlements.

The ecological and cultural transformations of the Holocene set the stage for the development of more complex societies in the British Isles. The interplay between rising sea levels, reforestation, and changing subsistence strategies illustrates the dynamic relationship between humans and their environment during this period. These changes not only shaped the landscapes and ecosystems of the British Isles but also laid the foundation for the agricultural and technological advancements that would define the Neolithic period and beyond. The story of post-glacial Britain is a testament to the resilience and adaptability of life in the face of profound environmental change, offering valuable lessons for navigating the challenges of the present and future.

Connections to Neighboring European Glaciations

The glaciation of Great Britain was not an isolated event but part of the broader network of glacial activity that shaped much of Europe during the Pleistocene Epoch. The Scandinavian and Alpine glaciations, in particular, were closely interconnected with the British glacial system. Massive ice sheets originating in Scandinavia advanced across the North Sea basin, merging with ice emanating from Scotland and northern England. This created a vast, continuous expanse of glacial ice that extended over large portions of northern Europe, including the British Isles. The interconnection of these ice sheets significantly influenced regional climates, hydrological systems, and ecosystems, creating a shared glacial heritage that linked Britain with its European neighbors (Clark et al., 2012).

One of the most significant effects of this interconnected glaciation was its influence on regional climate patterns. The vast ice sheets acted as massive heat sinks, cooling the surrounding atmosphere and altering weather systems across Europe. This cooling effect, coupled with changes in oceanic circulation caused by the influx of freshwater from melting glaciers, contributed to widespread climatic shifts that impacted the continent as a whole. Additionally, the presence of these ice sheets reshaped river systems and drainage patterns, redirecting the flow of major rivers such as the Rhine and Thames, which were often blocked or diverted by advancing glaciers (Benn & Evans, 2010).

During periods of lower sea levels, caused by the immense quantities of water locked in ice sheets, Britain and mainland Europe were connected by extensive land bridges. One of the most prominent of these was Doggerland, a fertile lowland area that spanned much of the southern North Sea. This land bridge facilitated the exchange of flora, fauna, and human populations, playing a crucial role in the migration and genetic diversity of early humans. For example, species such as red deer and aurochs moved freely between Britain and the continent, ensuring a shared gene pool and enhancing biodiversity. Similarly, human communities used these connections to migrate, trade, and share cultural practices, creating a network of interaction that shaped the development of societies across Europe (Gaffney et al., 2009).

The retreat of glaciers and subsequent rise in sea levels during the Holocene severed these connections, isolating Britain as an island. This isolation had profound ecological and cultural consequences. The ecosystems of Britain began to evolve independently, leading to the development of distinct flora and fauna. Certain species that were common across Europe became extinct in Britain due to their inability to recolonize after the land bridges were submerged, while others thrived in the newly isolated environment. For human populations, separation from continental Europe necessitated adaptations to island living, influencing trade, resource management, and societal development (Coles, 2002).

The shared impact of glaciation across Europe underscores the importance of studying these processes on a continental scale. The interactions between the British and European glacial systems reveal critical insights into the dynamics of ice sheets and their far-reaching effects. For instance, understanding how ice sheets merged and behaved collectively can shed light on mechanisms of ice flow and retreat, while examining the climatic changes associated with glaciation provides valuable context for modern climate studies. Additionally, the resilience of ecosystems during and after glaciation offers lessons in adaptation and recovery that are highly relevant to understanding the ecological impacts of present-day climate change (Clark et al., 2012; Benn & Evans, 2010).

The glaciation of Great Britain, deeply intertwined with broader European glacial systems, was a transformative force that shaped the geography, ecosystems, and human history of the region. From the vast ice sheets that connected Britain to mainland Europe to the eventual isolation that fostered distinct ecological and cultural identities, the legacy of the ice age is etched into the landscape and the story of its people. This shared glacial heritage highlights the interconnectedness of natural systems and the resilience of life in the face of profound environmental changes. By studying these past events, we not only gain a deeper understanding of the processes that have shaped our world but also draw valuable lessons for navigating the challenges of a rapidly changing climate. The enduring marks of glaciation on the British Isles remind us of the power of adaptation and the necessity of learning from Earth's dynamic history to guide us toward a sustainable future.

Chapter 6

Frozen Giants of North America

North American glaciation during the Pleistocene Epoch was one of the most significant climatic and geological events in the continent's history. Dominated by the massive Laurentide and Cordilleran Ice Sheets, this glaciation transformed North America's landscapes, ecosystems, and human history. This chapter explores the extent and dynamics of these ice sheets, their profound impact on geology and hydrology, their effects on flora, fauna, and human migration, and the implications of their retreat for understanding future climate changes.

Overview of the Laurentide and Cordilleran Ice Sheets

The Laurentide and Cordilleran Ice Sheets were the dominant ice masses during the North American glaciation, together covering an immense portion of the continent and profoundly influencing its geology, climate, and ecosystems. The Laurentide Ice Sheet, the larger of the two, originated over the Canadian Shield, a region of ancient, exposed bedrock that provided an ideal foundation for the accumulation of glacial ice. At its peak during the Last Glacial Maximum (LGM) approximately 20,000 years ago, the Laurentide Ice Sheet extended southward into the northern United States, reaching as far as Illinois, Ohio, and New York. Spanning roughly 13 million square kilometers, this ice sheet was a defining force in the transformation of North America, carving valleys, shaping river systems, and creating many of the continent's iconic landscapes (Dyke et al., 2002).

To the west, the Cordilleran Ice Sheet blanketed much of present-day British Columbia, Alaska, and parts of the northwestern United States. Although smaller than its Laurentide counterpart, the Cordilleran Ice Sheet played a pivotal role in shaping the rugged terrain of the western mountain ranges. Its extensive coverage of the Pacific Coastal Mountains, the Interior Plateau, and the Canadian Rockies left behind dramatic features such as deep fjords, steep-sided valleys, and sharp mountain ridges. The Cordilleran Ice Sheet's dynamics were heavily influenced by the region's topography, with glaciers funneling through narrow valleys and spreading out across plateaus. These processes created unique glacial landscapes, including the iconic cirques, moraines, and hanging valleys that are still visible today (Clague & Ward, 2011).

The interactions between the Laurentide and Cordilleran Ice Sheets were particularly significant in regions where they converged, such as the western Canadian Arctic and the Canadian Rockies. In these zones, their combined mass created dynamic glacial systems, influencing not only the local landscapes but also regional climates. For example, the combined ice sheets acted as massive reflectors of solar radiation, cooling the surrounding atmosphere and altering weather patterns across North America. This cooling effect extended far beyond the immediate vicinity of the ice, impacting ecosystems and climate systems across the continent.

Together, the Laurentide and Cordilleran Ice Sheets blanketed much of North America, fundamentally altering its physical landscapes and ecological systems. The weight of the ice depressed the Earth's crust, leading to the formation of proglacial lakes and setting the stage for future isostatic rebound as the ice melted. Their immense size also disrupted hydrological systems, redirecting rivers and influencing the drainage patterns that persist today. These ice sheets played a central role in shaping North America's geological history, leaving a legacy that continues to define the continent's landscapes and ecosystems.

Impact on North America's Geology and Hydrology

The immense weight and movement of the Laurentide and Cordilleran Ice Sheets during the Pleistocene Epoch profoundly reshaped North America's geology and hydrology, leaving behind a legacy of glacial landforms that continue to define the continent's physical and ecological character. As these ice sheets advanced and retreated over thousands of years, they sculpted the landscape through erosion and deposition, creating features that remain integral to the continent's geography.

One of the most striking examples of this transformation is the formation of the Great Lakes. The Laurentide Ice Sheet, through its sheer mass and movement, carved deep basins into the underlying bedrock of the northern United States and southern Canada. These basins, originally created by the relentless grinding action of ice loaded with sediment and debris, were later filled with meltwater as the ice sheets retreated. The resulting Great Lakes—Superior, Michigan, Huron, Erie, and Ontario—now form the largest freshwater system on Earth. These lakes provide vital ecological habitats, freshwater resources, and transportation corridors, supporting millions of people and a diverse range of species (Eyles, 2012). The Great Lakes are not only a testament to the power of glaciation but also an enduring resource that underpins regional economies and ecosystems.

In addition to the Great Lakes, the ice sheets left behind a variety of other glacial landforms that highlight the dynamic processes of ice movement. Drumlins, streamlined hills composed of glacial till, are scattered across regions such as upstate New York and southern Ontario, indicating the direction of ice flow. Moraines, ridges of debris deposited at glacier margins, also serve as prominent markers of glacial activity. Terminal moraines, such as the Oak Ridges Moraine in Ontario, mark the furthest extent of ice advances and are vital for understanding the chronological history of glaciation. Kettle lakes, formed by the melting of isolated ice blocks left behind in glacial sediments, and eskers, sinuous ridges of sand and gravel deposited by subglacial meltwater streams, further illustrate the retreating ice sheets' influence. These features collectively offer a window into the dynamic interplay between glacial processes and the evolving landscape.

The retreat of the Laurentide and Cordilleran Ice Sheets also dramatically altered North America's hydrology, leading to the creation of massive proglacial lakes. One of the largest and most significant of these was Lake Agassiz, which formed in the depressions left by the melting Laurentide Ice Sheet. At its peak, Lake Agassiz covered an area larger than any modern freshwater lake, and its periodic drainage events had far-reaching consequences. Catastrophic outbursts of meltwater from Lake Agassiz, released through glacial meltwater channels, reshaped river systems such as the Mississippi and St. Lawrence Rivers, carving new pathways and altering their courses. These outflows also contributed vast quantities of freshwater to the oceans, disrupting thermohaline circulation and influencing global climate systems, including triggering abrupt climate events like the Younger Dryas (Teller et al., 2002).

These hydrological changes were not limited to large-scale events; they also had profound impacts on ecosystems and human settlement patterns. The shifting river systems and newly created lakes provided habitats for aquatic species and resources for early human populations, shaping migration and settlement patterns. Fertile soils deposited by glacial outwash plains enriched agricultural regions, while the altered waterways became vital for transportation and trade. The long-term influence of glaciation on North America's hydrology is evident in the extensive network of rivers, lakes, and wetlands that continue to play a critical role in sustaining biodiversity and human livelihoods.

The geological and hydrological legacy of the Laurentide and Cordilleran Ice Sheets underscores the transformative power of glaciation in shaping the physical and ecological landscapes of North America. From the iconic Great Lakes to the dynamic river systems and proglacial lakes, these features serve as enduring reminders of the continent's glacial history and its ongoing influence on ecosystems and human development. This legacy also provides valuable insights into the long-term interactions between ice, water, and climate, offering lessons for understanding the potential impacts of contemporary glacial melting in the context of global climate change.

Effects on Flora, Fauna, and Early Human Migration

The glaciation of North America profoundly shaped its flora, fauna, and the migration patterns of early human populations, creating a dynamic and challenging environment that spurred significant evolutionary, ecological, and cultural adaptations. The advance of the Laurentide and Cordilleran Ice Sheets displaced vast ecosystems, compressing habitats into smaller refugia and forcing plant and animal species to adapt, migrate, or face extinction. Entire regions of the continent were rendered uninhabitable as glaciers covered up to 16 million square kilometers of land, while periglacial zones south of the ice sheets supported tundra-like vegetation. These harsh landscapes were dominated by hardy pioneer species such as mosses, lichens, and cold-adapted grasses, which stabilized soils and formed the basis for sparse ecosystems capable of surviving extreme conditions.

In these marginal environments, megafauna flourished, particularly in the open, cold-steppe habitats south of the glaciers. Iconic species such as woolly mammoths, mastodons, and saber-toothed cats thrived due to their unique adaptations to the cold. Thick fur, large body sizes to retain heat, and substantial fat reserves allowed these animals to endure freezing temperatures and forage in resource-scarce environments. Predatory species like the saber-toothed cat evolved powerful limbs and specialized dentition to hunt large herbivores, creating a dynamic predator-prey system that shaped these ecosystems. However, the retreat of the ice sheets and the accompanying warming climate dramatically altered these environments. As tundra landscapes gave way to forests, many megafaunal species, unable to adapt to rapid habitat changes and declining food resources, went extinct by the end of the Pleistocene (Faith & Surovell, 2009).

The retreat of the glaciers was a transformative period for North America's flora and fauna, as it opened pathways for the recolonization of northern landscapes and allowed ecosystems to diversify. Deciduous forests gradually replaced tundra in temperate regions, while coniferous forests expanded in colder northern zones. The newly reforested landscapes created a mosaic of habitats that supported a variety of species, including elk, moose, and beavers. These animals, along with numerous bird species and small mammals, recolonized the deglaciated areas, fostering the biodiversity seen today. Aquatic ecosystems also expanded with the formation of new lakes and rivers from glacial meltwater, providing critical habitats for fish, amphibians, and other freshwater species.

For early humans, the glaciation of North America posed significant challenges but also created unique opportunities. During the peak of glaciation, the Laurentide and Cordilleran Ice Sheets formed substantial physical barriers to human movement, confining populations to unglaciated areas in the south. These regions, including parts of modern-day Mexico and the southern United States, served as refugia where human populations adapted to the harsh glacial climate. However, the retreat of the ice sheets and the exposure of the Bering Land Bridge—a land connection between Siberia and Alaska— enabled one of the most significant migrations in human history. Around 15,000–20,000 years ago, early humans began crossing this land bridge into North America, dispersing rapidly across the continent and adapting to its diverse environments (Goebel et al., 2008).

Archaeological evidence, such as the Clovis culture, illustrates the ingenuity of these early inhabitants. Clovis tools, known for their distinctive fluted points, were designed for efficient hunting of large game, enabling humans to exploit megafaunal resources effectively. These tools represent a significant technological advancement, reflecting the adaptability and resourcefulness of early humans in meeting the challenges of their environment. As the climate warmed and ecosystems diversified, human populations adapted by broadening their subsistence strategies. Hunting shifted from a focus on megafauna to include smaller game, fishing, and the gathering of plant resources, such as nuts, seeds, and fruits. These changes in subsistence patterns not only reflect environmental adaptations but also indicate the emergence of more complex and regionally varied cultural practices.

The glaciation and subsequent deglaciation of North America were pivotal in shaping the continent's biological and cultural history. They forced life to evolve and adapt to extreme conditions, resulting in the extinction of some species, the migration and flourishing of others, and the emergence of human societies capable of thriving in diverse and challenging environments. These events highlight the resilience and adaptability of life and underscore the profound interconnectedness of climate, ecosystems, and human history.

Retreat of the Ice and Its Implications for Future Climates

The retreat of the Laurentide and Cordilleran Ice Sheets, beginning approximately 19,000 years ago, marked a turning point in North America's geological and ecological history, as the end of the Last Glacial Maximum (LGM) triggered profound transformations across the continent. Rising global temperatures initiated the melting of these massive ice sheets, releasing vast amounts of freshwater into oceans and profoundly disrupting the Earth's climate systems. The influx of freshwater from melting glaciers altered global thermohaline circulation, a system of ocean currents driven by temperature and salinity differences—resulting in cascading effects on climate. One of the most notable outcomes of these disruptions was the Younger Dryas, a period of sudden and temporary cooling that occurred approximately 12,900–11,700 years ago. This abrupt climatic reversal, potentially triggered by the release of meltwater from proglacial lakes into the North Atlantic, underscores the sensitivity of the Earth's climate to changes in ice sheet dynamics (Carlson, 2010).

The retreating glaciers not only affected global climate systems but also reshaped North America's hydrology and topography in profound ways. As the Laurentide and Cordilleran Ice Sheets receded, they left behind a legacy of proglacial lakes, formed in the depressions and basins created by the ice's immense weight. One of the most significant of these was Lake Agassiz, a colossal body of freshwater that at its peak covered an area larger than all of the modern Great Lakes combined. The periodic and often catastrophic drainage of Lake Agassiz released massive volumes of water into river systems such as the Mississippi and into the oceans. These outburst floods reshaped river courses, carved new valleys, and deposited vast amounts of sediment across the landscape, transforming the continent's hydrology.

These glacial meltwater floods also contributed to rising sea levels, inundating low-lying coastal areas and reshaping shorelines. The effects were not confined to North America; the addition of freshwater to the oceans had far-reaching global implications. By disrupting ocean currents, these events influenced weather patterns and monsoon systems, demonstrating the interconnectedness of ice, oceans, and atmospheric systems. The altered hydrology not only reshaped ecosystems but also created opportunities for human settlement and migration. Fertile floodplains and newly formed lakes provided abundant resources, encouraging the growth of human populations in these transformed landscapes.

The retreat of the ice sheets also exposed vast swaths of land that had been buried under ice for thousands of years, enabling the recolonization of these areas by flora and fauna. As the climate warmed, tundra vegetation gave way to forests, beginning with pioneer species such as birch and willow, followed by more complex deciduous and coniferous forests. These newly forested landscapes created diverse habitats that supported the expansion of animal populations, including elk, bison, and beavers, which thrived in the post-glacial environment. The changing ecosystems also had profound implications for early human populations, who adapted their subsistence strategies to exploit the new resources provided by the warming climate and reshaped terrain.

Studying the retreat of the Laurentide and Cordilleran Ice Sheets provides critical insights into the dynamics of ice sheets and their interactions with climate systems. Modern research uses these historical events as analogs to understand the ongoing impacts of contemporary glacial melting caused by anthropogenic climate change. Today, polar and alpine glaciers are retreating at unprecedented rates, contributing to rising sea levels, altering freshwater systems, and disrupting global climate patterns. The lessons from the Pleistocene highlight the interconnectedness of ice sheets, ocean circulation, and atmospheric systems in shaping Earth's climate. They also underscore the potential consequences of rapid glacial melting, including abrupt climate changes and widespread ecological and social impacts (Dyke et al., 2002).

The retreat of the Laurentide and Cordilleran Ice Sheets offers a valuable framework for understanding the complex feedback mechanisms between ice, climate, and hydrology. These historical events remind us that the Earth's systems are intricately linked and that changes in one component can trigger far-reaching consequences. As modern society grapples with the challenges of a warming planet, the study of these past glaciations emphasizes the importance of reducing human impact on the climate and preparing for the transformative changes that may come.

Enduring Legacy of North American Glaciation

The glaciation of North America during the Pleistocene Epoch left an indelible mark on the continent, profoundly shaping its landscapes, ecosystems, and human history. The immense power of the Laurentide and Cordilleran Ice Sheets carved the Great Lakes, rerouted rivers, and created fertile soils that continue to sustain life today. These ice sheets also drove dramatic ecological shifts, from the extinction of megafauna to the emergence of new ecosystems and the migration and adaptation of early human populations. As the ice retreated, it reshaped not only the geography of North America but also its climatic and hydrological systems, with impacts that resonate into the present.

The legacy of this glaciation extends beyond its physical and ecological transformations, serving as a powerful reminder of the dynamic nature of Earth's climate and the interconnectedness of its systems. Understanding the processes and consequences of past glacial cycles provides critical insights into the challenges of a warming planet. As we confront modern climate change, the history of North American glaciation equips us with the knowledge needed to anticipate, adapt to, and navigate the shifting landscapes of our future.

Chapter 7

Frozen Frontiers of South America

South America during the Pleistocene Epoch was a transformative force that shaped the continent's landscapes, ecosystems, and hydrological systems. While less extensive than glaciations in the Northern Hemisphere, South American glaciation displayed unique characteristics influenced by its geographical location, topography, and climatic conditions. This chapter explores the distinctive features of South American glaciation, its impact on the Andes Mountains and adjacent regions, its role in shaping river systems and ecosystems, and the evidence preserved in areas like Patagonia.

Unique Characteristics of South American Glaciation

South American glaciation was characterized by unique geographical and climatic conditions that distinguished it from the expansive ice sheets of the Northern Hemisphere, such as the Laurentide and Cordilleran. Unlike these vast, continuous glacial systems, South American glaciation was more localized, occurring predominantly in high-altitude regions of the Andes Mountains and the southernmost parts of the continent, particularly in Patagonia. This localization was driven by a combination of altitude and latitude, with the towering

69

Andes and the cool southern latitudes playing a central role in shaping glacial dynamics (Clapperton, 1993).

The Andes Mountains, stretching over 7,000 kilometers along the western edge of South America, served as the primary locus of glacial activity. High-altitude ice caps in the Andes generated extensive valley glaciers that flowed downward into lower elevations, carving dramatic landscapes along their paths. These glaciers were particularly prominent in regions like Peru, Bolivia, and northern Chile, where high elevations provided the necessary conditions for ice accumulation despite their location in tropical and subtropical latitudes. The Quelccaya Ice Cap in Peru, for example, remains one of the largest tropical ice caps in the world, preserving evidence of past glaciation that extends back thousands of years.

The Patagonian Ice Sheet, located in the southernmost part of South America, was the most significant glacial system on the continent. During the Last Glacial Maximum (LGM), approximately 20,000 years ago, the Patagonian Ice Sheet extended over an area of about 480,000 square kilometers, covering much of modern-day southern Chile and Argentina (Hulton et al., 2002). This ice sheet was responsible for shaping some of the continent's most iconic landscapes, including the fjords, U-shaped valleys, and glacial lakes that define the Patagonian region today. Notable features such as the Perito Moreno Glacier and the striking peaks of Torres del Paine stand as enduring testaments to the power of glaciation in this region.

The climatic conditions that supported glaciation in South America were significantly influenced by the proximity of the Southern Ocean and the Antarctic Circumpolar Current. The Southern Ocean acts as a massive thermal regulator, cooling the air masses that flow toward the continent. The Antarctic Circumpolar Current, the world's largest and most powerful ocean current, transports cold, nutrient-rich waters around Antarctica and into the southern latitudes of South America. This cold, moist air interacts with the towering Andes, creating orographic precipitation as moist winds are forced upward by the mountain range, leading to heavy snowfall and the accumulation of glacial ice in southern latitudes. This interaction was particularly pronounced in Patagonia, where the combination of latitude, elevation, and moisture from the Pacific Ocean created ideal conditions for the growth and maintenance of the Patagonian Ice Sheet.

In addition to latitude and altitude, South American glaciation was shaped by regional climatic variability associated with phenomena such as the El Niño-Southern Oscillation (ENSO). ENSO events influence precipitation and temperature patterns across the continent, periodically affecting the growth and retreat of glaciers. During cooler, wetter periods associated with La Niña events, glaciers advanced due to increased snowfall and lower temperatures. Conversely, warmer, drier periods linked to El Niño events contributed to glacial retreat.

The distinct characteristics of South American glaciation underscore the complex interplay of geographic, atmospheric, and oceanic forces in shaping the continent's ice systems. While smaller in scale than their Northern Hemisphere counterparts, the glaciers of South America were no less impactful, carving landscapes, shaping ecosystems, and leaving a legacy that continues to influence the region's environment and climate today. This unique context provides valuable insights into the dynamics of glaciation in regions where altitude and proximity to oceanic influences play a dominant role.

Impact on the Andes Mountains and Adjacent Regions

The glaciers of South America profoundly reshaped the Andes Mountains, creating some of the most dramatic and visually striking landscapes on the continent while also influencing adjacent regions and ecosystems. Over millennia, the movement and erosion caused by glaciers sculpted the mountains into the iconic features visible today, including U-shaped valleys, cirques, and jagged ridges. These landforms remain defining characteristics of Andean topography and serve as enduring evidence of the powerful forces of glaciation. Notable examples include the Fitz Roy Massif and Torres del Paine in

Patagonia, whose sharp, towering peaks and steep-walled valleys were carved by the relentless advance and retreat of Pleistocene glaciers. These features are not only geological wonders but also significant tourist destinations, showcasing the aesthetic and scientific value of glacially sculpted landscapes.

Glaciers also left their mark beyond the Andes themselves, profoundly influencing the hydrology of surrounding regions. Meltwater from the glaciers contributed to the formation of large proglacial lakes, such as Lake General Carrera and Lake Buenos Aires, which span the Chilean-Argentine border. These lakes were created when meltwater became trapped behind natural dams formed by moraines—ridges of debris deposited at glacier margins. The size and persistence of these lakes underscore the scale of glacial activity in the region and highlight their lasting influence on the local hydrological system. Even today, these lakes play a vital role in regulating water flow, supporting aquatic ecosystems, and providing water for human use and agricultural irrigation.

The sediments deposited by glaciers further transformed the surrounding regions. As glaciers advanced and retreated, they transported and deposited vast amounts of material, creating moraines, outwash plains, and till deposits. These sedimentary features enriched the soils of adjacent lowlands, making them fertile and suitable for vegetation. The deposition of nutrient-rich sediments was especially significant in areas like the Patagonian steppe, where the resulting soils supported diverse plant communities and became vital habitats for grazing animals such as guanacos. These fertile conditions also had implications for human populations, who later relied on these areas for agriculture and livestock grazing.

In addition to enriching soils, the glaciers influenced the biodiversity of the Andes and surrounding plains by creating dynamic and varied habitats. The glacial outwash plains, composed of gravel and sand carried by meltwater streams, provided critical habitats for specialized plant and animal species. These plains, characterized by sparse vegetation and seasonal water availability, became home to hardy plants adapted to nutrient-poor, rocky soils and to animals capable of surviving in harsh conditions. Over time, as vegetation became established on deglaciated terrain, these areas became focal points for ecological succession, promoting biodiversity and creating opportunities for new species to colonize the landscape.

The influence of South American glaciers extended well into the foothills and plains of the Andes, where meltwater-fed rivers and streams contributed to the development of wetlands and other unique ecosystems. These wetlands became hotspots of biodiversity, supporting a wide array of species, from migratory birds to aquatic plants and fish. The interconnectedness of these habitats illustrates the far-reaching impact of glaciation, demonstrating how ice-driven processes shaped not only the towering mountains but also the ecosystems and human settlements far beyond the glaciers' immediate reach.

Through their sculpting of the Andes Mountains and the transformation of adjacent regions, South American glaciers have left an enduring legacy. Their influence on hydrology, soil fertility, and biodiversity underscores the interconnectedness of glacial processes and ecosystems. These glacially shaped landscapes and habitats continue to play a vital role in the region's environmental and cultural identity, serving as both natural resources and powerful reminders of Earth's dynamic climatic history.

Glaciation's Role in Shaping South American River Systems and Ecosystems

The glaciation of South America played a pivotal role in sculpting the continent's river systems and ecosystems, creating dynamic hydrological networks and fostering biodiversity that persists to this day. The advance and retreat of glaciers over millennia dramatically altered the course and structure of major rivers, especially in Patagonia, where the interaction between ice and water has left a lasting imprint on the landscape. Rivers such as the Rio Baker and Rio Santa Cruz, two of the largest in the region, owe their origins and current configurations to the processes of glaciation. During the Last Glacial Maximum (LGM), these rivers were fed by extensive meltwater from retreating glaciers, which carved valleys, created new channels, and redirected existing flows. As the ice retreated, braided river systems formed, characterized by multiple intertwining channels that shifted seasonally, depositing sediment and creating a mosaic of aquatic habitats (Sugden et al., 2005). These braided systems not only reshaped the geography of Patagonia but also became vital ecological zones, supporting a diverse array of plant and animal life.

Glacial retreat also facilitated the formation of deltas and wetlands, particularly in areas where meltwater streams reached the ocean or inland basins. These regions became rich in biodiversity, offering critical habitats for migratory birds, fish, and amphibians. For instance, the delta of the Rio Santa Cruz provides a sanctuary for a wide range of species, while the wetlands of southern Patagonia serve as breeding grounds for birds such as flamingos and waterfowl. These environments illustrate the interconnectedness of glacial processes and ecological systems, demonstrating how the retreat of ice can create new opportunities for life to flourish.

One of the most significant contributions of glacially influenced rivers was their role in sediment transport. As glaciers advanced and retreated, they ground down bedrock into fine sediment known as glacial flour. This sediment, carried downstream by meltwater, deposited in riverbeds, floodplains, and coastal zones, played a critical role in shaping the physical and ecological characteristics of these environments. Glacial flour enriched aquatic ecosystems by providing essential nutrients that stimulated the growth of plankton and algae, forming the foundation of aquatic food chains. These nutrient-rich waters supported thriving populations of fish and invertebrates, which in turn sustained larger predators and human fishing communities. The sediment also contributed to the fertility of floodplains, allowing plant species to colonize and creating habitats for terrestrial animals.

The glacial legacy of South America extends beyond its hydrology to its biodiversity, as glaciation forced plant and animal species to adapt to dynamic and often harsh conditions. The repeated advance and retreat of glaciers created fragmented landscapes, pushing species to migrate, adapt, or find refugia in unglaciated areas. These refugia, such as isolated valleys and lowland basins, served as safe havens for species during glacial periods, allowing them to survive and eventually recolonize deglaciated regions during interglacial periods. These cycles of retreat and recolonization fostered genetic diversity and promoted the evolution of species uniquely adapted to the Andean environment.

Iconic species such as the Andean condor and guanaco exemplify the resilience and adaptability required to thrive in glacially influenced ecosystems. The Andean condor, with its impressive wingspan and soaring capabilities, is well-suited to navigate the rugged terrains and open skies of the Andes. Similarly, the guanaco, a relative of the llama, is adapted to the arid steppe and mountainous regions shaped by glacial processes. These species, along with many others, highlight the intricate relationships between glaciation and biodiversity in South America.

Glaciation also influenced the distribution and diversity of plant life. As glaciers retreated, they exposed bare, nutrient-poor landscapes that were gradually colonized by pioneer species such as mosses, lichens, and grasses. These plants stabilized the soil and paved the way for the establishment of more complex ecosystems, including shrublands and forests. The Andes, in particular, became a hotspot of plant diversity, with species adapting to the varied microclimates created by glacial landforms. Today, the Andean region is recognized as one of the world's most biodiverse mountain ranges, with ecosystems ranging from tropical cloud forests to high-altitude paramos.

The glacial imprint on South America's river systems and ecosystems underscores the profound interconnectedness of geological and biological processes. The dynamic interplay between ice, water, and life not only shaped the physical landscape but also fostered the resilience and diversity of the continent's ecosystems. This legacy continues to influence the region, providing a valuable framework for understanding how natural systems respond to climatic changes, past and present.

Evidence from Patagonia and Other Areas

Patagonia, located at the southernmost tip of South America, is a region rich with evidence of past glaciation, offering a comprehensive record of the dynamic processes that shaped the continent during the Pleistocene Epoch. Dominated by the Patagonian Ice Sheet during the Last Glacial Maximum (LGM), this area experienced extensive glacial activity that left indelible marks on its landscapes. The Patagonian Ice Sheet, which spanned much of southern Chile and Argentina, reached its maximum extent approximately 20,000 years ago and has since retreated, revealing a wealth of geological and sedimentary features that document its history.

Physical Landforms: Evidence of Glacial Movement

The physical landforms of Patagonia provide some of the most direct evidence of glacial activity. Terminal moraines, such as those found in the Lago Argentino basin, delineate the maximum extent of glacial advances and serve as natural markers of the ice sheet's progression. These ridges of debris, left behind as glaciers reached their furthest limits, provide invaluable data about the size and timing of glacial events. Additionally, polished bedrock and striations—linear grooves carved into the rock by debris-laden ice—trace the paths of glacial movement and reveal the immense erosive power of the ice sheet. Features such as U-shaped valleys, fjords, and hanging valleys further underscore the transformative impact of glaciation on Patagonia's rugged terrain.

Prominent examples of glacially sculpted landscapes in Patagonia include the stunning fjords along the Chilean coast and the iconic Perito Moreno Glacier. These features are not only geological wonders but also active systems that continue to demonstrate glacial processes in real time. The fjords, carved by glaciers advancing into the ocean, now serve as important habitats for marine life and are vital for understanding the interactions between ice, land, and sea.

Beyond its striking landforms, Patagonia also holds a rich archive of sedimentary records that offer insights into the climatic and glacial history of the region. Sediment cores extracted from lakes and fjords, such as Lago Argentino and Seno Última Esperanza, contain finely layered deposits that were transported and laid down by glacial meltwater. These layers, known as varves, provide a chronological record of glacial activity, with thicker layers corresponding to periods of increased melting and sedimentation.

Isotopic analyses of these sediments have been particularly revealing. By examining oxygen and carbon isotopes, researchers have been able to infer changes in temperature and precipitation over thousands of years. Such studies have illuminated the timing of glacial advances and retreats and have shown how these changes correspond to global climatic events such as the Younger Dryas and Holocene warming periods (Heusser, 2003). These findings not only enhance our understanding of regional climate dynamics but also contribute to broader studies of Earth's climate systems.

Further evidence of South American glaciation comes from ice cores extracted from high-altitude glaciers in the Andes, such as the Quelccaya Ice Cap in Peru. These cores preserve a wealth of information about past climatic conditions, including temperature, precipitation, and atmospheric composition. By analyzing trapped air bubbles, researchers can reconstruct levels of greenhouse gases such as carbon dioxide and methane, providing critical data about the interplay between glacial cycles and atmospheric chemistry.

Ice cores from Quelccaya and similar sites have revealed records spanning thousands of years, documenting not only local climate changes but also global phenomena such as the Little Ice Age and the Medieval Warm Period. These records are particularly valuable for understanding the interconnectedness of Earth's climate systems, as they show how changes in the Southern Hemisphere were linked to events in the Northern Hemisphere, such as shifts in thermohaline circulation and variations in solar radiation.

Evidence of past glaciation in Patagonia and the Andes has far-reaching implications for both science and society. The glacial features and sedimentary records of this region provide a critical baseline for understanding how ice sheets respond to climate change, offering analogs for the current retreat of glaciers in the context of global warming. Modern observations of retreating glaciers in Patagonia highlight the urgency of addressing climate change, as these ice masses are key sources of freshwater and play a vital role in regional ecosystems.

The interconnectedness of the geological, climatic, and ecological systems in Patagonia serves as a reminder of the complex relationships that govern Earth's environment. As researchers continue to uncover the secrets of Patagonia's glacial history, the region remains a vital area for studying the past, present, and future dynamics of our planet's climate.

Glaciation of South America, particularly in Patagonia and the Andes, stands as a testament to the transformative power of ice in shaping landscapes, ecosystems, and climate systems. From the towering peaks and sculpted valleys of the Andes to the sedimentary archives preserved in glacial lakes and fjords, the legacy of these glaciations offers a window into Earth's dynamic past. These features not only reveal the interplay between regional and global climatic forces but also highlight the resilience of life in the face of profound environmental change. As modern glaciers retreat under the pressures of anthropogenic climate change, the lessons of South American glaciation become ever more relevant. By understanding the forces that shaped this region, we gain critical insights into the interconnectedness of Earth's systems and the potential consequences of our warming planet. The story of South American glaciation reminds us of the importance of preserving these natural archives as we navigate an uncertain environmental future.

Chapter 8

Breathing Spaces Between Ice Ages

The interglacial periods, representing relatively brief intervals of warmth between prolonged glacial stages, have played a pivotal role in shaping Earth's landscapes, ecosystems, and human history. These "breathers" between ice ages are characterized by significant climatic, geological, and biological transformations that continue to influence our planet's evolutionary trajectory. This chapter explores the definition and importance of interglacial, their geological and biological markers, notable interglacial periods in Earth's history, and their profound impacts on biodiversity and human development.

Definition and Significance of Interglacial

Interglacial are critical periods of climatic and ecological transformation that occur between the colder and longer glacial phases within a glacial cycle. Characterized by warmer global temperatures, these intervals are marked by the retreat of massive ice sheets, the rise of sea levels, and the resurgence of ecosystems into areas that were previously dominated by ice. During interglacials, temperate and tropical conditions expand poleward, drastically altering landscapes and promoting biodiversity. The primary drivers of these periods are variations in Earth's orbit, axial tilt, and precession—collectively referred to as Milankovitch cycles. These orbital variations influence the amount and distribution of solar radiation received by the planet, triggering shifts in global temperatures and glacial dynamics (Hays et al., 1976).

Interglacials stand in stark contrast to glacial periods, which can persist for tens of thousands of years and are characterized by widespread glaciation, colder temperatures, and lower sea levels. In comparison, interglacials are relatively brief, lasting approximately 10,000–20,000 years. Despite their brevity, these warm periods play a pivotal role in Earth's climate system by resetting and reshaping ecosystems and hydrological systems and allowing for evolutionary and cultural progress.

Transformative Effects on Earth's Systems

The significance of interglacial lies in their ability to transform Earth's systems in profound and lasting ways. These warm intervals provide crucial opportunities for ecosystems to recover from the harsh, resource-scarce conditions of glacial periods. As ice sheets retreat, vast tracts of land are exposed, creating new habitats for plants and animals. This re-expansion of ecosystems enables species to recolonize deglaciated areas, adapt to changing conditions, and diversify. For example, during the Eemian interglacial (~130,000–115,000 years ago), boreal forests expanded northward into regions that were previously uninhabitable, supporting a surge in biodiversity and creating new ecological networks (Dutton & Lambeck, 2012).

Interglacial also play a critical role in the hydrological cycle. Rising global temperatures during these periods accelerate the melting of ice sheets, contributing to higher sea levels and altering ocean circulation patterns. Meltwater from retreating glaciers and ice caps feeds into rivers, lakes, and oceans, reshaping aquatic ecosystems and providing vital freshwater resources for terrestrial life. The redistribution of water during interglacials influences weather patterns, precipitation, and the availability of resources, further driving ecological and environmental change.

Impact on Human Development

Interglacials have had a profound impact on human history, providing the stable climatic conditions necessary for key periods of cultural and technological advancement. The Holocene, which began approximately 11,700 years ago, serves as a prime example of an interglacial period that has shaped human civilization. The relatively stable climate of the Holocene has allowed humans to transition from nomadic hunter-gatherer societies to agricultural and urbanized civilizations. The advent of agriculture during this period enabled the development of surplus food production, population growth, and the establishment of complex social structures.

The predictability of seasonal weather patterns during interglacial has also facilitated human innovation. From early irrigation systems to advancements in architecture and infrastructure, the stable conditions of interglacial have provided a foundation for human ingenuity to flourish. In addition, the warming temperatures and melting ice have revealed fertile lands and accessible waterways, further encouraging exploration, trade, and cultural exchange.

Importance for Understanding Earth's Past and Future

Studying interglacial is essential for reconstructing Earth's climatic history and for predicting future changes in the context of global warming. Interglacial provide critical baselines for understanding the natural variability of Earth's climate and for distinguishing between natural and anthropogenic influences. By examining the dynamics of past interglacial, such as the Eemian and the Holocene, scientists can gain insights into how ecosystems, sea levels, and atmospheric conditions respond to warming temperatures.

Moreover, interglacial offer valuable analogs for understanding the potential impacts of ongoing climate change. The current Holocene interglacial, now influenced by human activities, is characterized by unprecedented levels of greenhouse gas concentrations, which are pushing global temperatures beyond the natural bounds of previous interglacial. This anthropogenic warming poses significant risks to biodiversity, sea levels, and human societies, underscoring the importance of mitigating climate change while adapting to its inevitable impacts.

Interglacial, though fleeting in geological time, are periods of immense transformation and opportunity. They highlight the interconnectedness of Earth's systems and the delicate balance required to sustain life. As we navigate the challenges of a rapidly changing climate, the lessons of interglacial remind us of the resilience and adaptability of natural systems and the need to preserve the stability that has enabled human prosperity.

Geological and Biological Markers of Interglacial

Interglacial periods are identified and studied through a combination of geological and biological markers, which serve as windows into Earth's climatic past. These markers provide invaluable data on changes in temperature, precipitation, atmospheric composition, and ecosystem dynamics, allowing scientists to piece together a detailed picture of interglacial environments and their impacts on the planet.

Geological Markers: A Record of Environmental Change

Geological markers form the foundation of interglacial studies, offering physical evidence of past climatic conditions preserved in sedimentary deposits, ice cores, and isotopic signatures. **Ice cores** from Greenland and Antarctica are among the most critical resources for understanding interglacial periods. These cores contain layers of compacted snow that have accumulated over millennia, trapping air bubbles that preserve ancient atmospheric gases. The analysis of these gases, such as carbon dioxide (CO_2) and methane (CH_4), reveals elevated levels during interglacial, consistent with warmer global temperatures. For example, ice cores from Antarctica have provided direct evidence of interglacial warmth during the Eemian period (~130,000–115,000 years ago), showing CO_2 levels exceeding 280 parts per million (ppm), which is significantly higher than levels observed during glacial periods (Petit et al., 1999).

Marine sediment cores also play a pivotal role in identifying interglacial. These cores, extracted from ocean floors, contain layers of sediment deposited over time. Variations in the ratio of oxygen isotopes ($16O^{16}O16O$ and $18O^{18}O18O$) in the shells of marine organisms provide insights into past ocean temperatures and global ice volume. During interglacial, lower concentrations of $18O^{18}O18O$ in seawater reflect reduced ice volumes and warmer ocean temperatures, offering a clear signal of these warm intervals. These isotopic records align with other geological evidence to construct timelines of interglacial events.

In addition, sedimentary deposits on land, such as lake varves and loess deposits, serve as geological markers of interglacial periods. Lake varves, composed of annual layers of sediment, provide high-resolution records of seasonal changes in temperature and precipitation. Loess deposits, formed by wind-blown dust during glacial periods, often contain intercalated layers of soil that developed during interglacial, reflecting warmer and wetter conditions conducive to soil formation.

Biological Markers: Traces of Life in a Changing Climate

Biological markers complement geological evidence by documenting changes in ecosystems and species distributions during interglacial. Fossilized pollen is a particularly valuable tool for reconstructing past vegetation patterns. During interglacial periods, pollen from temperate forest species, such as oak, beech, and birch, replaces pollen from tundra plants like grasses and sedges, indicating a shift to warmer climates. These pollen records are often preserved in sediment cores from lakes and peat bogs, offering a continuous record of vegetation changes over time.

The presence of species that thrive in temperate climates further supports the identification of interglacial. For example, the spread of deciduous forests into higher latitudes during interglacial periods is a hallmark of these warm intervals. Similarly, the proliferation of certain mollusks and coral reefs in marine environments indicates warmer ocean temperatures and stable sea levels. Coral reef growth, in particular, is a reliable indicator of interglacial conditions, as these ecosystems require specific temperature and salinity ranges to thrive. Fossilized coral reefs, such as those from the Eemian interglacial, provide direct evidence of elevated sea levels and warmer oceans during these periods.

Biological markers also include faunal evidence, such as the migration and range expansion of animal species. During interglacial periods, many species migrate poleward in response to warming temperatures, leaving behind fossilized remains that document these shifts. For instance, the presence of temperate-adapted mammals, such as red deer and wild boar, in regions previously dominated by cold-adapted species signals interglacial conditions. These migrations not only highlight the adaptability of species but also underscore the dynamic nature of ecosystems during these periods.

Integrating Geological and Biological Markers

The integration of geological and biological markers enables scientists to construct detailed and accurate reconstructions of interglacial periods. By combining evidence from ice cores, marine sediments, and terrestrial fossils, researchers can identify the timing, duration, and environmental characteristics of interglacial. For example, the synchronous rise in atmospheric CO_2 levels (geological marker) and the spread of temperate forests (biological marker) provides a robust indicator of interglacial onset.

This multidisciplinary approach also sheds light on the mechanisms driving interglacial climates. The interplay between solar radiation, greenhouse gas concentrations, and feedback processes—such as changes in albedo due to ice melt—is reflected in both geological and biological records. Together, these markers provide a comprehensive understanding of the factors that define interglacial and their role in Earth's climate system.

Key Interglacial Periods in Earth's History

Throughout Earth's climatic history, several interglacial periods stand out for their profound influence on the planet's physical landscapes, ecosystems, and the trajectory of human evolution and development. These intervals, though brief in geological terms, represent critical chapters in the story of Earth's climate, offering valuable insights into the dynamics of warm periods, their drivers, and their impacts on the biosphere and climate systems.

Eemian Interglacial

Marine Isotope Stage 5e, ~130,000–115,000 Years Ago

The Eemian Interglacial, also known as Marine Isotope Stage 5e, was one of the warmest interglacial of the last 500,000 years. Global temperatures during this period were approximately 1–2°C warmer than today, with significant reductions in polar ice volumes. These conditions led to a sea level rise of 5–9 meters above current levels, flooding low-lying coastal regions and creating new marine habitats (Dutton & Lambeck, 2012). The Eemian's warm climate allowed boreal forests to expand into higher latitudes, replacing tundra in regions such as Scandinavia and Siberia. This period of ecological expansion supported diverse megafaunal populations, including woolly mammoths and saber-toothed cats, which thrived in the rich ecosystems of Europe, Asia, and North America.

The Eemian is particularly significant for its parallels to contemporary climate conditions, as it provides a natural analog for understanding the impacts of elevated greenhouse gas levels and reduced ice cover. The high-resolution records from Eemian ice cores and marine sediments reveal how small changes in solar radiation, amplified by feedback mechanisms, can lead to dramatic shifts in climate. Studying the Eemian also underscores the potential vulnerability of ice sheets to sustained warming, with implications for modern sea-level rise and coastal resilience.

The Holocene

~11,700 Years Ago to Present

The Holocene, Earth's current interglacial period, has been instrumental in shaping human history and civilization. Beginning approximately 11,700 years ago, the Holocene has been characterized by relatively stable climate conditions, with global temperatures remaining within a narrow range. This stability has facilitated the development of agriculture, urbanization, and technological innovation, making the Holocene the foundation of modern society.

During the early Holocene, warming temperatures triggered the retreat of glaciers and the expansion of ecosystems into previously glaciated regions. Forests replaced tundra in many parts of the Northern Hemisphere, creating fertile environments for plant and animal life. This climatic shift coincided with the Neolithic Revolution, as humans transitioned from nomadic lifestyles to settled agricultural societies. Key civilizations, including those in Mesopotamia, the Indus Valley, and China, emerged during this period, taking advantage of the reliable growing seasons and abundant resources provided by the Holocene climate.

The Holocene has also seen significant human impacts on the environment, particularly during the Anthropocene, a proposed epoch marking the profound influence of human activity on Earth's systems. Deforestation, habitat fragmentation, and rising greenhouse gas emissions have altered the natural dynamics of the Holocene, pushing the climate toward unprecedented conditions. Understanding the Holocene's stability and its disruptions is critical for addressing contemporary challenges related to climate change, biodiversity loss, and sustainable development.

Pliocene-Quaternary Transition

~2.58 Million Years Ago

The transition into the Quaternary Period marked a pivotal moment in Earth's climatic history, introducing the glacial-interglacial cycles that have defined the past 2.58 million years. Before this transition, during the relatively stable and warm conditions of the Pliocene Epoch, Earth's climate was characterized by smaller temperature fluctuations and the absence of large-scale glaciations. However, the onset of cyclic glaciations during the Quaternary, driven by Milankovitch cycles—periodic changes in Earth's orbit, axial tilt, and precession—ushered in a new era of dramatic climatic variability. This era saw alternating glacial and interglacial phases, profoundly reshaping global landscapes, ecosystems, and the evolutionary trajectories of life.

The "First Interglacial" of the Quaternary, one of the earliest warm intervals following the onset of glacial cycles, created an environment that fostered the diversification and expansion of early hominins. As ice sheets retreated and temperatures rose, forests and grasslands spread across deglaciated regions, offering abundant resources and new ecological niches. These changes likely played a crucial role in the evolution of species within the genus *Homo*, driving adaptations to varying environments and resource availability. The expansion of grasslands, for instance, may have encouraged early hominins to develop bipedal locomotion, a key evolutionary milestone that allowed for more efficient movement across open terrains.

This transition also influenced the migration and distribution of species. The warmer and more stable conditions of the "First Interglacial" allowed plants and animals to colonize newly available habitats, promoting biodiversity and the development of complex ecosystems. This reshaping of the landscape not only affected flora and fauna but also set the stage for early human innovations in tool use, social organization, and subsistence strategies, as hominins adapted to their changing environments.

The Pliocene-Quaternary transition underscores the intricate interplay between climate shifts and evolutionary processes. The cyclical nature of glacial and interglacial periods created alternating pressures and opportunities for life, driving adaptation, resilience, and innovation. These cycles also highlight the profound impact of climate on the Earth system, as each interglacial provided a unique set of conditions that shaped the planet's physical and biological history.

In addition to its evolutionary implications, this transition offers valuable insights into the mechanisms driving Earth's climatic variability. By studying the geological and biological records from this period, scientists can better understand the natural drivers of glacial and interglacial phases, providing context for the unprecedented changes occurring in today's climate. The Pliocene-Quaternary transition serves as a reminder of the dynamic nature of Earth's climate and its profound influence on the development of life. As we face the challenges of contemporary climate change, the lessons of this transition underscore the importance of adaptation, resilience, and the interconnectedness of Earth's systems.

Marine Isotope Stage 11

~400,000 Years Ago

Marine Isotope Stage (MIS) 11 is one of the most remarkable interglacial periods in Earth's Quaternary history, often referred to as a "super-interglacial" due to its exceptional duration and stability. Spanning approximately 30,000 years, from roughly 424,000 to 374,000 years ago, MIS 11 was characterized by global temperatures comparable to or slightly warmer than those of the pre-industrial Holocene. This extended period of climatic stability allowed ecosystems to thrive, sea levels to rise significantly, and polar ice sheets to retreat extensively. Sea levels during MIS 11 were estimated to be 6–13 meters higher than today, underscoring the sensitivity of ice sheets to even modest temperature increases (Berger & Loutre, 2003).

The prolonged stability of MIS 11 created ideal conditions for ecosystems to flourish and adapt over extended timeframes. In terrestrial environments, forests expanded into higher latitudes, replacing tundra and creating diverse habitats that supported a wide range of plant and animal species. The warming temperatures and increased precipitation associated with MIS 11 fostered the development of temperate and boreal forests across Europe, Asia, and North America. These forests provided critical habitats for megafaunal species such as mammoths, mastodons, and saber-toothed cats, as well as for smaller animals and migratory birds.

In marine environments, the warmer seas and higher sea levels during MIS 11 facilitated the growth of coral reefs and other marine ecosystems, which flourished in the stable climate. Coastal regions, transformed by rising seas, became hotspots of biodiversity as marine life adapted to the changing habitats. The sustained warmth and ecological stability of MIS 11 highlight the interconnectedness of climate, biodiversity, and habitat evolution.

One of the most significant aspects of MIS 11 is its orbital configuration, which closely resembles that of the current Holocene interglacial period. Both periods are marked by low eccentricity in Earth's orbit, resulting in relatively stable levels of solar radiation. This orbital configuration is believed to have contributed to the unusually long duration of MIS 11, as it minimized the variations in solar forcing that typically drive transitions between glacial and interglacial phases. By studying MIS 11, scientists can gain insights into the natural variability of interglacial climates and the factors that influence their longevity.

MIS 11 has been extensively studied for its relevance to contemporary and future climate conditions. As the most analogous interglacial period to the Holocene, it provides a valuable baseline for understanding the potential impacts of current human-driven climate change. For example, the higher sea levels during MIS 11 underscore the vulnerability of polar ice sheets to prolonged warmth, offering important lessons for predicting the response of the Greenland and Antarctic ice sheets to ongoing warming.

Furthermore, the stability of MIS 11 highlights the resilience of Earth's climate system under natural conditions, while also drawing attention to the unprecedented rate of change caused by anthropogenic influences. The study of MIS 11 emphasizes the importance of maintaining ice sheet stability to mitigate long-term sea-level rise, as even modest changes in temperature can have profound and lasting impacts on global coastlines and ecosystems.

The lessons of MIS 11 extend beyond its immediate parallels to the Holocene. Its duration and stability demonstrate how natural feedback mechanisms, such as the interplay between ice sheet dynamics, ocean circulation, and atmospheric CO_2 levels, can sustain warm periods for tens of thousands of years. However, the conditions that led to MIS 11's stability are unlikely to recur in the near future due to the rapid and unprecedented increase in greenhouse gas concentrations caused by human activities.

Understanding MIS 11 helps contextualize the long-term trends of interglacial periods and provides a framework for assessing the impacts of modern climate change. By comparing the natural stability of MIS 11 with the accelerated warming of the Anthropocene, scientists can better anticipate the challenges of ice sheet loss, sea-level rise, and ecosystem disruptions in the coming centuries.

Implications of Interglacial Periods

Each interglacial period has played a distinct and transformative role in Earth's climate history, reshaping landscapes, ecosystems, and evolutionary pathways in ways that continue to influence the planet today. These periods of warmth have served as crucial intervals of recovery and renewal, allowing ecosystems to rebound from the harsh conditions of glacial stages. During interglacials, the retreat of massive ice sheets has exposed land, reshaped river systems, and created new habitats, fostering biodiversity and enabling species to adapt, evolve, and expand their ranges. These climatic transitions have not only sculpted the physical geography of the planet but also driven the development of complex ecosystems and the evolutionary trajectories of life on Earth.

Interglacials have also provided critical opportunities for human adaptation and innovation. For example, the stable and temperate conditions of the Holocene allowed early humans to transition from hunter-gatherer lifestyles to settled agricultural societies, sparking the rise of civilizations. Similarly, interglacial periods like the Eemian enabled the proliferation of early human populations, who took advantage of the expanding forests, fertile lands, and abundant resources created by warmer climates. These periods highlight the intricate relationship between climate and human development, illustrating how environmental stability can foster technological and cultural advancements.

By examining interglacial periods, scientists gain valuable insights into the mechanisms that drive climate transitions and the resilience of Earth's systems in the face of change. Through the study of ice cores, sediment layers, and biological markers, researchers can reconstruct detailed records of past interglacial, revealing the complex interplay between solar radiation, greenhouse gases, ocean circulation, and feedback processes. This understanding is essential for contextualizing current climate trends, as it allows scientists to distinguish natural variability from anthropogenic impacts.

Interglacial periods also serve as critical benchmarks for understanding Earth's climatic future, particularly in the context of ongoing global warming. The Holocene, for instance, has been significantly altered by human activities, pushing atmospheric carbon dioxide levels far beyond the natural range observed in previous interglacial. This anthropogenic warming has profound implications for ice sheet stability, sea-level rise, and ecosystem resilience, drawing direct parallels to the dramatic changes that occurred during past interglacial periods. Studying these historical intervals provides a valuable framework for predicting and mitigating the potential impacts of contemporary climate change.

As the planet faces unprecedented challenges from human-driven warming, the lessons of interglacial underscore the importance of maintaining the delicate balance that supports life on Earth. These periods remind us of the resilience of natural systems, the interconnectedness of climate and biodiversity, and the critical need for sustainable practices to preserve the stability that has enabled human prosperity. By understanding the transformative power of interglacial, we can better anticipate the challenges ahead and develop strategies to navigate the complex relationship between human activity and the planet's climate systems.

Impacts on Biodiversity and Human Development

Interglacial periods have profound and multifaceted impacts on biodiversity, ecosystems, and human development, serving as critical windows of opportunity for species to thrive and adapt. These warm intervals, characterized by retreating ice sheets and rising global temperatures, reshape landscapes and create new ecological niches, allowing for the expansion, evolution, and diversification of life on Earth.

The retreat of massive ice sheets during interglacial periods exposes vast expanses of land that were previously locked under ice. These newly available areas provide fertile ground for ecosystems to recolonize and flourish. Temperate forests, grasslands, and wetlands often establish themselves in these deglaciated regions, creating habitats that support a wide variety of species. This ecological resurgence promotes genetic diversity and resilience within populations, as species adapt to the warming conditions and expand their geographic ranges.

For example, during the Eemian interglacial (~130,000–115,000 years ago), warming conditions allowed coral reefs to flourish at higher latitudes than their current range, expanding marine biodiversity and providing critical habitats for a variety of marine organisms. Similarly, terrestrial species benefited from the expansion of boreal and temperate forests, which replaced tundra landscapes and created environments rich in resources and opportunities for interaction. In addition to fostering biodiversity, interglacial periods provide a backdrop for evolutionary innovation. Species that migrate into newly available habitats often encounter different environmental pressures, leading to speciation and the emergence of new ecological roles. These processes underscore the importance of interglacials as drivers of long-term evolutionary change.

While interglacial periods create opportunities, they also present significant ecological challenges. The rapid climatic changes that accompany transitions from glacial to interglacial conditions can result in the extinction of species unable to adapt to new environments or shifting ecological dynamics. Megafauna, such as woolly mammoths and saber-toothed cats, often faced dwindling habitats and food resources as ecosystems transformed, leading to their eventual extinction during these transitional phases. These extinctions highlight the vulnerability of species to rapid environmental change, even during periods of warming and recovery.

Human activities during the Holocene, Earth's most recent interglacial, have exacerbated the ecological challenges associated with interglacials. Habitat destruction, deforestation, and the overexploitation of natural resources have accelerated biodiversity loss, compounding the stresses placed on ecosystems by natural climatic shifts. Anthropogenic impacts have fundamentally altered the balance of interglacial systems, transforming them into periods of intensified environmental pressure.

Interglacial periods have also played a crucial role in shaping human history, providing the stability needed for cultural and technological advancements. Warmer climates and abundant resources during the Holocene, for instance, facilitated the development of agriculture, the establishment of permanent settlements, and the rise of complex civilizations. The stable climatic conditions of interglacials allowed humans to innovate, expand their populations, and modify their environments to an unprecedented extent, laying the groundwork for modern societies.

Understanding the role of interglacials in human and ecological history is essential for addressing contemporary challenges. The impacts of anthropogenic climate change, including rising temperatures, habitat destruction, and biodiversity loss, threaten to destabilize the conditions that have supported human prosperity and ecological balance. Interglacials remind us of the interconnectedness of Earth's systems and the delicate equilibrium required to sustain life. By learning from the natural processes and challenges of past interglacials, we can better navigate the complexities of our rapidly changing world and work toward preserving the stability and diversity that define these critical intervals.

Interglacial periods are pivotal intervals in Earth's climatic and ecological history, offering moments of transformation and opportunity amid the cycles of glaciation. These warm phases have not only reshaped landscapes and ecosystems but also provided the stability necessary for profound evolutionary advancements and the rise of human civilization. Yet, they also remind us of the delicate balance that sustains life on Earth, as their transitions often bring challenges such as species extinctions and environmental disruptions. As we grapple with the unprecedented pace and scale of anthropogenic climate change, the lessons of interglacials become increasingly relevant. They underscore the interconnectedness of Earth's systems, the resilience of life, and the consequences of destabilizing natural processes. By understanding and respecting the dynamics of interglacials, we can better prepare to adapt, innovate, and act as stewards of a planet whose history is intricately tied to these critical windows of warmth and renewal.

Chapter 9

Last Cold Kingdoms in Northern Europe

The glaciations of Northern Europe, spanning several million years, have profoundly shaped the region's geography, ecosystems, and human history. These glacial periods, interspersed with warmer interglacial intervals, left indelible marks on the landscapes of Scandinavia, the Baltic region, and the British Isles. From the carving of fjords to the creation of fertile soils, the impacts of glaciation are seen in every aspect of the environment and have influenced the development of early European societies. This chapter explores the timeline and extent of glaciation across Northern Europe, its effects on specific regions, the long-term environmental consequences, and the ice's influence on cultural evolution.

Timeline and Extent of Glaciation Across Northern Europe

The glaciations of Northern Europe were defining events of the Quaternary Period, beginning approximately 2.58 million years ago and persisting into the present-day Holocene Epoch. These glaciations were not continuous but occurred in cycles, alternating between cold glacial periods and warmer interglacial intervals. This cyclic pattern was driven by Milankovitch cycles—periodic variations in Earth's orbit, axial tilt, and precession that affect the distribution and intensity of solar radiation, leading to significant shifts in global temperatures and ice coverage (Hays et al., 1976).

One of the earliest major glaciations in Northern Europe, the Elsterian glaciation (~400,000–300,000 years ago), marked the advance of extensive ice sheets over large areas, including present-day Germany, Poland, and the British Isles. The ice sheets during this period dramatically altered the landscapes they covered, scouring the surface and leaving behind features such as glacial troughs and ridges (Ehlers & Gibbard, 2007). As the Elsterian ice retreated, it set the stage for subsequent glaciations that would further reshape the region.

The Saalian glaciation (~300,000–130,000 years ago) represented an even more expansive phase of glaciation, with ice sheets extending deep into Central Europe. This glaciation significantly modified the topography of Northern Europe, carving deep valleys, depositing thick layers of till, and influencing river courses. Notably, the Saalian ice sheets contributed to the formation of extensive proglacial lakes, as meltwater became trapped by advancing and retreating ice margins (Ehlers & Gibbard, 2007).

The most recent and extensively studied glaciation, the Weichselian (known as the Devensian in the British Isles), began approximately 115,000 years ago and reached its zenith during the Last Glacial Maximum (LGM) around 20,000 years ago. At its peak, the Weichselian glaciation covered vast swathes of Northern Europe. The Fennoscandian Ice Sheet, one of the dominant ice masses, spanned millions of square kilometers, covering most of Scandinavia, the Baltic region, and parts of the British Isles. This immense ice sheet extended southward, reaching as far as northern Germany and Poland (Ehlers & Gibbard, 2007; Benn & Evans, 2010).

The Fennoscandian Ice Sheet was remarkable not only for its size but also for its thickness, which reached up to 3 kilometers in central Scandinavia. The immense weight and movement of this ice sheet profoundly altered the physical landscapes beneath it. Glacial erosion sculpted U-shaped valleys, fjords, and cirques, particularly in mountainous regions like Norway. The ice sheet also deposited vast amounts of sediment, forming moraines, eskers, and drumlins that continue to shape the topography of the region (Benn & Evans, 2010).

In the Baltic region, the retreat of the Weichselian ice sheets resulted in the formation of the Baltic Sea, a feature that originated from proglacial lakes fed by meltwater (Ehlers et al., 2011). In the British Isles, the same glaciation shaped iconic features such as the Lake District, Glen Coe, and the English Channel, which was exposed during periods of lower sea levels caused by the trapping of water in continental ice sheets (Clark et al., 2012).

The cumulative impact of these glaciations extended far beyond their immediate geological effects. By carving valleys, creating basins for lakes, and redistributing sediments, these glaciations fundamentally altered Northern Europe's hydrology, soils, and ecosystems. The legacy of these glaciations is evident in the region's diverse landscapes, which bear the marks of ice age processes and continue to influence human activity and ecological systems today.

Effects on Scandinavia, the Baltic Region, and the British Isles

The glaciation of Scandinavia created some of the world's most breathtaking and iconic landscapes, a testament to the power of glacial processes. The immense movement and weight of the Fennoscandian Ice Sheet during the Quaternary sculpted dramatic fjords along the Norwegian coastline. Fjords such as the Sognefjord, which stretches over 200 kilometers and plunges to depths exceeding 1,300 meters, are among the deepest and longest fjords globally, showcasing the erosive capabilities of glacial activity (Benn & Evans, 2010). These fjords, carved by glacial tongues advancing through mountain valleys, are now filled with seawater and remain central to Norway's natural beauty and cultural identity.

Glacial erosion also played a key role in shaping Scandinavia's valleys and mountain ranges, leaving behind U-shaped valleys, steep-walled cirques, and prominent moraines. These features are characteristic of glacially sculpted terrains, with U-shaped valleys such as Jotunheimen in Norway standing as a stark contrast to the V-shaped valleys formed by river erosion. As the Fennoscandian Ice Sheet retreated, it deposited vast amounts of till—unsorted glacial sediment—across the landscape, creating fertile soils that support agriculture today. The retreat also gave rise to thousands of lakes, including Sweden's Lake Vänern, the largest in the European Union, and Lake Vättern. These lakes are essential for biodiversity, freshwater resources, and human use.

In the Baltic region, glaciation profoundly influenced the region's topography and hydrology. The Baltic Sea, a defining feature of Northern Europe, owes its origin to glacial processes. As the ice sheets melted, the resulting proglacial lakes merged and filled basins scoured by glacial erosion, eventually giving rise to the Baltic Sea. The region is also characterized by abundant glacial deposits, including eskers—long, sinuous ridges of sand and gravel deposited by meltwater rivers beneath glaciers—and drumlins, elongated hills of glacial till. These features are widespread across Finland, Estonia, and Latvia, creating a patchwork of fertile plains and rolling hills. These glacial formations not only shape the agricultural landscape but also influence water management practices, as the region's numerous rivers and lakes are often fed by glacially carved basins.

The British Isles were significantly impacted by glaciation, particularly during the Devensian glaciation, the last major glacial advance in the region. Ice sheets enveloped much of Scotland, Ireland, and northern England, carving dramatic valleys and reshaping the topography. Glen Coe in Scotland is a prime example of a U-shaped valley formed by glacial erosion, with its steep sides and broad floor illustrating the immense power of ice movement. Other iconic features, such as the Lake District, owe their rugged beauty and countless lakes to glacial sculpting and deposition. The region's landscapes, dotted with moraines, tarns, and drumlins, remain a testament to the glacial processes that once dominated.

The retreat of ice sheets during the Devensian glaciation had profound effects on the geography of the British Isles. One of the most significant changes was the exposure of the English Channel, which separated Britain from mainland Europe. During periods of lower sea levels, such as the Last Glacial Maximum, Britain was connected to the continent via the land bridge known as Doggerland. As the ice melted and sea levels rose, this land bridge was submerged, isolating the British Isles and shaping their cultural and ecological evolution. This separation influenced migration patterns, the spread of flora and fauna, and the development of unique ecosystems and human societies in the region.

Across Scandinavia, the Baltic region, and the British Isles, glaciation left a legacy of striking physical features, fertile landscapes, and dynamic water systems. These glacial processes not only sculpted the natural environment but also laid the foundation for human settlement, agriculture, and cultural development in Northern Europe.

Long-Term Environmental Impacts

The glaciations of Northern Europe have had enduring environmental impacts, particularly on the region's soil and water systems, leaving a legacy that continues to shape its landscapes and ecosystems. As the massive ice sheets advanced and retreated over millennia, they profoundly altered the land, stripping surface materials and reshaping topographies.

The abrasive action of advancing ice sheets scoured the land, removing vegetation, soil, and loose rock while exposing the underlying bedrock. This process resulted in thin, nutrient-poor soils in many parts of Northern Europe, particularly in areas like Scandinavia and northern Scotland, where rocky and rugged landscapes now dominate (Benn & Evans, 2010). These regions are often characterized by sparse vegetation and limited agricultural potential. However, the retreat of the glaciers also deposited vast quantities of sediment, such as till and loess, in other areas. In regions like the Baltic lowlands, eastern

England, and parts of Denmark, these deposits enriched the soil, creating fertile agricultural lands. The loess deposits, fine windblown sediments often associated with glacial margins, are particularly important for farming, supporting crops and contributing to the economic sustainability of these areas (Ehlers & Gibbard, 2007).

Glacial meltwater systems played a pivotal role in shaping the hydrology of Northern Europe. As ice sheets melted, proglacial lakes formed at their edges, often covering vast areas and acting as reservoirs for glacial meltwater. One notable example is the Baltic Ice Lake, a massive freshwater body that existed during the retreat of the Fennoscandian Ice Sheet. Periodic outburst floods from such lakes dramatically reshaped river systems, carving new channels and altering existing ones. The Rhine and Thames river systems, for instance, were significantly influenced by these catastrophic flood events (Ehlers et al., 2011). These changes not only modified the physical geography of the region but also created rich, dynamic habitats for aquatic species and laid the groundwork for the modern riverine landscapes that support agriculture, transport, and water supply.

The legacy of glacial meltwater systems is evident in the numerous rivers, lakes, and wetlands that characterize Northern Europe today. Many of the region's lakes, such as those found in Sweden and Finland, are remnants of proglacial or ice-dammed lakes, offering biodiversity-rich habitats for a variety of species. These water systems also provide critical resources for human populations, including drinking water, hydroelectric power, and recreational opportunities (Benn & Evans, 2010).

In addition to its effects on soil and water systems, the removal of the massive weight of glacial ice has triggered isostatic rebound, the gradual rise of land as it adjusts to the loss of the overlying ice. This process has been particularly pronounced in Scandinavia, where the land around the Gulf of Bothnia is rising by several centimeters per century (Ehlers & Gibbard, 2007). Isostatic rebound has reshaped coastlines, creating new landforms such as raised beaches, coastal terraces, and emergent islands. These changes continue to influence regional sea levels and coastal ecosystems, presenting both opportunities and challenges for human activity. For example, areas experiencing rapid uplift may benefit from expanded land for development or agriculture, but they also face disruptions to existing coastal infrastructure.

The interplay of these glacial legacies—soil formation, hydrological transformations, and isostatic rebound—has had cascading effects on biodiversity and human settlement. Fertile soils in previously glaciated regions have supported agricultural development for millennia, while the extensive river and lake systems have served as hubs of biodiversity and economic activity. Understanding these enduring impacts provides valuable insights into the ways in which past glaciations continue to shape modern landscapes and ecosystems, highlighting the intricate connections between geological processes and human-environment interactions.

Ice's Influence on Early European Societies

The glaciation of Northern Europe not only transformed its physical landscapes but also profoundly shaped the cultural evolution of early human societies. The challenges posed by harsh glacial climates forced humans to develop innovative survival strategies, laying the groundwork for significant technological and social advancements. During glacial periods, much of Northern Europe was uninhabitable due to extensive ice cover, and human populations were confined to unglaciated refugia in southern Europe. In these refugia, early humans honed critical skills to cope with cold climates, such as advancements in tool-making, clothing, and shelter construction (Fagan, 2011). Stone tools became more specialized, with finely crafted blades and points adapted for hunting large game in periglacial environments. Additionally, the development of insulated clothing from animal hides and the construction of durable shelters allowed early humans to survive in frigid conditions, demonstrating remarkable ingenuity and resilience.

As interglacial periods brought warmer climates and retreating ice sheets, human populations migrated northward, capitalizing on the newly deglaciated landscapes. These migrations were driven by the opportunities presented by fertile soils, abundant freshwater, and the expansion of forests and grasslands created by glacial processes. In Northern Europe, including the British Isles, the development of agriculture marked a transformative phase in human history. The fertile soils deposited by glaciers, enriched with loess and till, supported the cultivation of crops and the domestication of animals, fostering the rise of permanent settlements. Early Neolithic communities in regions such as eastern England and Scotland established farming practices, enabling the transition from hunter-gatherer lifestyles to agrarian societies (Roberts, 2014). These agricultural advancements laid the foundation for complex societies, facilitating trade, innovation, and the growth of early European civilizations.

The influence of glacial landscapes extended beyond practical adaptations to cultural identity and mythology. The dramatic fjords of Scandinavia, carved by the Fennoscandian Ice Sheet, became enduring symbols of the natural world's power and beauty. Central to Norse mythology, these landscapes were often associated with divine forces, inspiring stories that celebrated the interplay between humans and the environment. Similarly, the rugged terrains of the British Isles, shaped by glacial activity, inspired rich folklore and cultural narratives. Features such as glacial valleys, craggy peaks, and ancient stone circles embedded in glacially influenced landscapes often found their way into legends and myths, reflecting humanity's deep connection to and reverence for the natural world (Armstrong, 2009).

The legacy of Northern Europe's glaciation also continues to influence cultural and economic practices. Modern societies in glaciated regions still utilize the fertile soils, freshwater resources, and dramatic landscapes created by glacial processes. The enduring cultural significance of these landscapes, seen in art, literature, and tourism, highlights the lasting impact of glaciation on both the environment and the human experience. Understanding the interplay between glacial processes and cultural evolution offers valuable insights into how humans have adapted to and been inspired by the dynamic forces of nature throughout history.

Chapter 10

Ice to Inferno of the Greenhouse Planet

The greenhouse effect is a natural phenomenon essential for maintaining life on Earth, but human activities have amplified this process, pushing the planet into an unprecedented era of warming. This chapter explores the mechanisms of the greenhouse effect, historical examples of natural greenhouse periods, the role of the Industrial Revolution in intensifying the effect, and the current and future challenges posed by anthropogenic climate change.

Explanation of the Greenhouse Effect and Its Drivers

The greenhouse effect is a critical natural process that enables life on Earth by maintaining a habitable temperature range. It operates through the interaction between incoming solar radiation, Earth's surface, and specific gases in the atmosphere. Solar radiation passes through the atmosphere and warms the Earth's surface, which, in turn, emits energy in the form of infrared radiation. Instead of allowing this energy to escape into space, greenhouse gases (GHGs)—including carbon dioxide (CO_2), methane (CH_4), water vapor (H_2O), nitrous oxide (N_2O), and ozone (O_3)—absorb and re-radiate the heat, effectively insulating the planet. This process keeps Earth's average surface temperature at approximately 15°C, compared to an estimated -18°C if the greenhouse effect did not exist (IPCC, 2021). Without this natural phenomenon, Earth would be too cold to sustain most forms of life, making the greenhouse effect indispensable to the planet's habitability.

Natural drivers of the greenhouse effect have regulated Earth's climate over millions of years, maintaining a delicate balance of atmospheric GHG concentrations. Volcanic activity is one such driver, periodically releasing large amounts of CO_2 into the atmosphere during eruptions. These emissions have contributed to long-term climatic shifts, such as periods of global warming during the Mesozoic Era. Similarly, wetlands, permafrost, and oceans are natural sources of methane, a potent greenhouse gas with a global warming potential more than 25 times greater than that of CO_2 over a 100-year period (Berner, 2004). Water vapor, the most abundant GHG, amplifies the greenhouse effect through a feedback mechanism: as temperatures rise, more water evaporates, further enhancing the warming process (Trenberth et al., 2009).

Despite the stability provided by natural processes, human activities have increasingly disrupted the greenhouse effect since the Industrial Revolution. The combustion of fossil fuels for energy, deforestation, and industrial processes have led to a dramatic increase in CO_2 levels. Atmospheric CO_2 concentrations, which remained below 300 parts per million (ppm) for thousands of years, have risen to over 420 ppm as of 2023 (NOAA, 2023). Methane emissions from agricultural activities, waste management, and energy production have also surged, compounding the warming effect. These anthropogenic contributions have intensified the greenhouse effect, driving rapid climatic changes that outpace the natural variability observed in Earth's geological history.

The amplification of the greenhouse effect due to human activity is evident in the acceleration of global warming. The rate of temperature increase since the late 19th century is unprecedented, with the global average temperature now approximately 1.2°C higher than pre-industrial levels (IPCC, 2021). This intensification is altering weather patterns, melting ice sheets, and increasing the frequency of extreme weather events, underscoring the profound impact of disrupted GHG balances on Earth's climate systems.

Historical Examples of Natural Greenhouse Periods

Throughout Earth's history, elevated greenhouse gas concentrations have driven significant warming events that profoundly influenced the planet's climate, ecosystems, and evolutionary pathways. These periods highlight the delicate interplay between atmospheric greenhouse gases and global climate systems, offering critical insights into the long term effects of increased carbon levels.

One of the most studied examples of a natural greenhouse period is the Paleocene-Eocene Thermal Maximum (PETM), approximately 56 million years ago. The PETM was triggered by a sudden and massive release of carbon into the atmosphere and oceans, with potential sources including extensive volcanic activity, the combustion of organic carbon stores, or the destabilization of methane hydrates on the ocean floor (Zachos et al., 2008). During this event, global temperatures rose by 5–8°C within a few thousand years—a rapid shift on geological timescales. This abrupt warming caused ocean acidification, leading to the extinction of many deep-sea species, while terrestrial ecosystems experienced significant biogeographic changes. Tropical and subtropical species migrated toward the poles, exploiting newly hospitable regions, while other species faced severe stress or extinction due to rapid environmental changes. The PETM serves as a stark example of the cascading ecological impacts that can result from elevated greenhouse gas concentrations.

Another prominent greenhouse period occurred during the Mesozoic Era, particularly in the Cretaceous Period (~145–66 million years ago). This era was marked by persistently high atmospheric CO_2 levels, largely driven by prolonged volcanic activity associated with the breakup of Pangaea and the formation of major ocean basins (Royer, 2014). CO_2 concentrations during the Cretaceous are estimated to have been several times higher than present-day levels, creating a warm, ice-free planet. Tropical conditions extended to polar regions, resulting in widespread lush vegetation, including forests in Antarctica. These greenhouse conditions supported diverse ecosystems, including the dominance of dinosaurs and the proliferation of marine reptiles, ammonites, and rudist reefs. However, the high CO_2 levels also led to a sluggish ocean circulation system, resulting in anoxic events that periodically disrupted marine ecosystems and caused mass extinctions.

Natural greenhouse periods are not confined to deep geological time. More recent examples include the interglacial periods within the Quaternary Period, such as the current Holocene Epoch, which began approximately 11,700 years ago. Interglacials are characterized by higher levels of atmospheric CO_2 and CH_4 compared to glacial periods, resulting in warmer global climates. During the Holocene, these greenhouse gas levels were augmented by natural processes, such as wetland methane emissions and the gradual release of CO_2 from thawing permafrost and ocean reservoirs (Petit et al., 1999). These elevated greenhouse gas levels provided the climatic stability that has allowed human civilizations to flourish, supporting the development of agriculture, permanent settlements, and complex societies.

These examples demonstrate the profound and varied impacts of elevated greenhouse gas levels on Earth's climate and ecosystems. They illustrate how shifts in atmospheric composition, whether through natural processes or human activities, can initiate widespread changes in temperature, precipitation patterns, and biodiversity. The PETM and Cretaceous periods reveal the potential for rapid warming to disrupt ecosystems, while interglacial periods highlight the role of greenhouse gases in creating climates conducive to life. Understanding these natural greenhouse periods offers valuable context for assessing the current anthropogenic intensification of the greenhouse effect and its implications for Earth's future.

Industrial Revolution Intensified the Greenhouse Effect

The Industrial Revolution, which began in the late 18th century, marked a pivotal moment in both human history and Earth's climate systems. This era of technological and economic transformation introduced a dependence on fossil fuels—coal, oil, and natural gas that fundamentally altered the composition of the atmosphere. The combustion of these fuels for energy, transportation, and industrial processes released massive amounts of carbon dioxide (CO_2), the most abundant anthropogenic greenhouse gas, into the atmosphere (Crutzen, 2002). Alongside CO_2, industrial activities also generated significant emissions of methane (CH_4) and nitrous oxide (N_2O), two other potent greenhouse gases that further intensified the greenhouse effect.

Since the pre-industrial era, the atmospheric concentration of CO_2 has risen from approximately 280 parts per million (ppm) to over 420 ppm as of 2023, a dramatic increase that coincides with the rapid expansion of fossil fuel use during and after the Industrial Revolution (NOAA, 2023). This unprecedented growth in CO_2 levels is primarily driven by the combustion of fossil fuels, which accounted for the majority of energy production during the 19th and 20th centuries and remains dominant in the 21st century. Industrial processes such as cement production have further contributed to CO_2 emissions by releasing stored carbon from geological materials.

Methane emissions have also surged, driven by agricultural activities, particularly livestock farming and rice cultivation. Livestock, such as cattle, produce methane as a byproduct of digestion, while rice paddies emit CH_4 due to anaerobic conditions in flooded fields. Methane emissions from landfills and the extraction of fossil fuels add to this burden. Although CH_4 is present in the atmosphere at much lower concentrations than CO_2, it has a global warming potential that is more than 25 times greater over a 100-year period, making it a critical driver of climate change (IPCC, 2021). Nitrous oxide, another potent greenhouse gas, is released through the use of synthetic fertilizers in agriculture and industrial processes, adding to the complex web of anthropogenic emissions.

Deforestation and land-use changes further exacerbate the greenhouse effect by reducing the Earth's capacity to absorb atmospheric CO_2 through photosynthesis. Forests, which serve as critical carbon sinks by storing carbon in biomass and soils, have been cleared at alarming rates to make way for agriculture, urbanization, and infrastructure development. The loss of these carbon sinks not only releases stored carbon into the atmosphere but also diminishes the planet's ability to sequester future emissions, creating a compounding feedback loop. Between 2001 and 2020, an estimated 411 million hectares of forest were lost globally, with significant contributions to atmospheric CO_2 levels (Global Forest Watch, 2021).

The Industrial Revolution not only initiated the rise in greenhouse gas emissions but also established modern industrial and economic systems heavily reliant on carbon-intensive practices. Factories powered by coal, oil-based transportation systems, and urban centers requiring vast amounts of energy became hallmarks of industrialization. These systems facilitated unprecedented economic growth and technological advancement but at the cost of accelerating climate change. The reliance on fossil fuels and deforestation during this period laid the foundation for the modern challenges associated with anthropogenic climate change.

The implications of this historical turning point are far-reaching. The elevated levels of greenhouse gases in the atmosphere have driven global warming, leading to observable changes in weather patterns, melting glaciers, rising sea levels, and the increased frequency of extreme weather events. As the legacy of the Industrial Revolution continues to shape the global climate, understanding its role in the intensification of the greenhouse effect is essential for developing strategies to mitigate its impacts and transition to a sustainable future.

Current Implications and Future Challenges

The intensification of the greenhouse effect has profound and far-reaching implications for the planet's climate, ecosystems, and human societies. Since the late 19th century, the global average temperature has risen by approximately 1.2°C, leading to a cascade of environmental and social challenges. More frequent and severe heatwaves, rising sea levels, and the increased intensity of storms and flooding are becoming hallmarks of a warming planet, with disproportionate impacts on vulnerable populations and ecosystems (IPCC, 2021). Polar ice sheets and glaciers, such as those in Greenland and Antarctica, are melting at accelerated rates, contributing significantly to rising sea levels. These changes pose existential threats to coastal communities, small island nations, and biodiversity-rich marine ecosystems.

Biodiversity is under severe strain, as many species struggle to adapt to the rapid pace of climate change. Coral reefs, often referred to as the rainforests of the ocean, are experiencing widespread bleaching events due to ocean warming and acidification caused by elevated CO_2 levels. These reefs support approximately 25% of all marine species, and their loss would ripple through global ecosystems, affecting fisheries, tourism, and coastal protection (Hoegh-Guldberg et al., 2018). On land, changing temperatures and shifting precipitation patterns are disrupting habitats, leading to migration, population declines, and in some cases, extinction. These disruptions threaten food security and water availability, particularly in regions already facing resource scarcity.

The future challenges posed by the intensified greenhouse effect are monumental. Without significant reductions in greenhouse gas (GHG) emissions, global temperatures are projected to rise by 2–4°C by 2100, exceeding critical climate tipping points (Steffen et al., 2018). These tipping points include the potential collapse of the West Antarctic Ice Sheet, the dieback of the Amazon rainforest, and the release of methane from thawing permafrost. Such changes would trigger irreversible impacts on global climate systems, leading to amplified warming and more extreme weather events. Additionally, sea-level rise could displace hundreds of millions of people by the end of the century, compounding humanitarian crises and straining global resources.

Mitigating the effects of the intensified greenhouse effect requires immediate and coordinated action. Transitioning to renewable energy sources, such as solar, wind, and geothermal, is essential to reducing dependence on fossil fuels. Innovations in carbon capture and storage technologies, along with reforestation and wetland restoration, can help sequester atmospheric carbon and slow the pace of warming. Protecting and expanding natural carbon sinks, such as forests and peatlands, is equally critical to maintaining the delicate balance of Earth's carbon cycle.

Adapting to the realities of a warmer planet is just as crucial as mitigation efforts. Building resilient infrastructure, especially in vulnerable coastal and urban areas, can help communities withstand rising sea levels and extreme weather events. Developing sustainable agricultural practices and enhancing early warning systems for disasters will be key to safeguarding food and water supplies. Global cooperation, informed by scientific understanding and guided by principles of equity and sustainability, is essential for addressing these interconnected challenges.

The intensified greenhouse effect represents one of the most urgent and complex challenges humanity has ever faced. It is a stark reminder of the interconnectedness of Earth's systems and the consequences of disrupting them. While the scale of the problem is immense, so too is the potential for collective action and innovation. By reducing emissions, restoring ecosystems, and preparing for the inevitable impacts of a warming planet, humanity can not only mitigate the worst outcomes but also build a more sustainable and resilient future. The history of life on Earth shows that adaptation and ingenuity are hallmarks of survival; in this new era of warming, these qualities will be our greatest assets. The choices we make today will determine the legacy we leave for generations to come, a planet capable of sustaining life in all its diversity or one irrevocably altered by inaction.

Chapter 11

When the World Chilled Again

The Little Ice Age (LIA), spanning from approximately 1300 to 1850 CE, represents a period of cooler global temperatures that left a profound imprint on Earth's climate, ecosystems, and human societies. This chapter explores the timeline and climatic characteristics of the Little Ice Age, examines its causes, discusses its significant impacts on agriculture, economies, and societies, and analyzes the recovery and transition to the modern warming period.

The Little Ice Age of 1300–1850 CE

The Little Ice Age (LIA) was a complex climatic event characterized by a series of cooling periods interspersed with brief warm intervals, rather than a uniformly cold era. Spanning roughly from the 14th to the 19th century, the LIA peaked during the 17th century, with global temperatures averaging 0.5°C–1°C lower than those of the early 20th century (Fagan, 2011; Mann et al., 2009). These modest-sounding temperature declines had disproportionately large impacts, as even slight reductions in global averages can lead to significant regional climatic changes.

Regional variations in the timing and intensity of cooling defined the Little Ice Age, with some areas experiencing particularly severe conditions. Europe and North America bore the brunt of these climatic shifts, enduring harsh winters, shortened growing seasons, and cooler summers. In Eastern Europe, rivers like the Thames in England and the Seine in France froze regularly during the winter months, events that were rare before and after this period. Similarly, in North America, indigenous populations had to adapt to shifting ecological conditions, as colder winters affected food sources and migration patterns of game animals.

Glacial activity during the LIA reached its zenith in the 17th century, leaving an indelible mark on many landscapes. In the European Alps, advancing glaciers inundated valleys, overrunning farmland and villages. Iconic glaciers such as the Rhône Glacier in Switzerland expanded dramatically, encroaching on human settlements and farmland, and causing widespread disruption (Grove, 2004). In Iceland and Greenland, sea ice extended further south, disrupting traditional fishing routes and making maritime trade increasingly perilous. These changes not only strained local economies but also contributed to the isolation of Norse settlements in Greenland, hastening their decline.

Evidence of the Little Ice Age's impact is preserved in multiple natural archives. Tree ring analysis offers valuable insights, as tree growth is highly sensitive to temperature and precipitation fluctuations. Narrow tree rings from this period indicate reduced growing seasons and cooler temperatures across much of the Northern Hemisphere. Similarly, ice core data from Greenland and Antarctica reveal reduced concentrations of greenhouse gases and isotopic markers consistent with cooler global climates. Sediment records from lakes and riverbeds also confirm evidence of increased erosion and glacial runoff, indicative of expanded ice cover and heightened glacial activity (Esper et al., 2002).

The persistence of cooler conditions during the Little Ice Age resulted in widespread ecological and societal stress. The reduced growing seasons impacted agricultural productivity, leading to food shortages, increased malnutrition, and in some cases, famine. These ecological stresses were not uniform, but they underscore how even modest shifts in average global temperatures can cascade into far-reaching consequences for natural systems and human societies.

Volcanic Activity, Solar Minima, and Oceanic Circulation Shifts

The Little Ice Age (LIA) is attributed to a combination of interconnected natural factors, including volcanic activity, reductions in solar radiation, and shifts in oceanic circulation patterns, all of which contributed to the prolonged period of cooler global temperatures. These factors, acting in concert, created feedback loops that exacerbated the cooling trends observed during this period.

Volcanic Activity

Volcanic eruptions played a significant role in initiating and amplifying the cooling trends of the Little Ice Age. Large eruptions inject massive amounts of sulfur dioxide (SO_2) and ash particles into the stratosphere, where they form sulfate aerosols. These aerosols reflect incoming solar radiation, reducing the amount of energy reaching Earth's surface and causing short-term cooling effects. Notable eruptions during the LIA include Mount Samalas in 1257, Huaynaputina in 1600, and Laki in 1783 (Oppenheimer, 2003). These eruptions were powerful enough to trigger temporary but significant cooling events, which, when compounded by other climatic factors, prolonged the overall cooling of the Little Ice Age. For example, the eruption of Mount Tambora in 1815 caused the "Year Without a Summer" in 1816, leading to widespread crop failures and societal disruptions.

Reduced Solar Radiation

Periods of reduced solar activity, such as the Spörer Minimum (1460–1550) and the Maunder Minimum (1645–1715), coincided with some of the coldest decades of the Little Ice Age. These solar minima are characterized by a significant reduction in sunspot activity, which correlates with lower solar irradiance. While the exact mechanisms linking solar minima to global cooling are complex, reduced solar output likely weakened the energy balance of Earth's climate system, amplifying the effects of other cooling factors (Eddy, 1976). Changes in solar radiation also impact atmospheric circulation patterns, potentially altering storm tracks and precipitation distribution, further contributing to the observed cooling.

Shifts in Oceanic Circulation Patterns

Changes in ocean circulation patterns, particularly in the North Atlantic, played a critical role in sustaining the cooler conditions of the Little Ice Age. The Atlantic Meridional Overturning Circulation (AMOC), a key component of global thermohaline circulation, may have weakened during this period, reducing the transport of warm water from the tropics to higher latitudes. This slowdown could have been triggered by an influx of freshwater from melting glaciers or increased sea ice, which disrupted the salinity-driven sinking of water in the North Atlantic (Rahmstorf et al., 2005). This disruption not only cooled the North Atlantic region but also affected global heat distribution, reinforcing the cooling trends of the Little Ice Age.

Additionally, the interaction between oceanic and atmospheric systems, such as the North Atlantic Oscillation (NAO), influenced regional climate variability during the LIA. A prolonged negative phase of the NAO, characterized by weaker westerly winds and colder winters in Europe, is thought to have been prevalent during parts of the Little Ice Age, exacerbating cooling in the Northern Hemisphere.

Interplay of Factors

The combination of these natural factors created a complex climate system that reinforced cooling. For example, volcanic activity could trigger short-term cooling, which, when coupled with reduced solar radiation and shifts in oceanic circulation, extended and amplified the cooling trends. These factors interacted with feedback mechanisms, such as increased snow and ice cover, which enhanced Earth's albedo (reflectivity), further reducing temperatures and sustaining the Little Ice Age.

The multifaceted causes of the Little Ice Age highlight the intricate interplay between Earth's natural systems and underscore the sensitivity of the climate to small perturbations. Understanding these factors provides valuable insights into the dynamics of Earth's climate and the potential implications of future climatic shifts.

Oceanic Circulation Shifts

Changes in ocean circulation patterns, particularly the slowing of the Atlantic Meridional Overturning Circulation (AMOC), were significant contributors to the prolonged cooler conditions during the Little Ice Age. The AMOC is a critical component of Earth's thermohaline circulation, which redistributes heat globally by moving warm, saline water from the tropics to the North Atlantic, where it cools, sinks, and flows back toward the equator at depth. During the Little Ice Age, cooler temperatures and increased freshwater input from melting glaciers and sea ice disrupted this delicate system, reducing the strength of the AMOC and altering heat distribution across the Northern Hemisphere (Rahmstorf et al., 2005).

The reduced AMOC weakened the transport of warm tropical waters to higher latitudes, leading to significant regional cooling in Europe and North America. This cooling effect was particularly pronounced in the North Atlantic, where surface waters became colder, amplifying the already reduced solar radiation and volcanic aerosol effects of the period. This disruption also caused changes in storm tracks and precipitation patterns, leading to harsher winters and altered weather systems across the Northern Hemisphere.

The influx of freshwater from melting ice and increased sea ice cover during the Little Ice Age further exacerbated the slowing of the AMOC. Freshwater is less dense than saline water, and its addition to the North Atlantic surface reduced the sinking of cold, salty water that drives thermohaline circulation. This feedback loop not only prolonged cooler conditions but also contributed to the formation of proglacial lakes and expanded sea ice coverage, both of which reflected more solar radiation and further cooled the region.

In addition to its regional impacts, the disruption of the AMOC likely had global repercussions. The altered heat distribution may have contributed to shifts in tropical and subtropical weather patterns, including changes in monsoon systems and prolonged droughts in parts of Africa and Asia. The weakened AMOC also influenced the upwelling of nutrient-rich waters in the Southern Ocean, affecting marine ecosystems and carbon cycling on a global scale (Broecker, 1997).

Overall, the slowing of the AMOC during the Little Ice Age underscores the interconnectedness of Earth's climate systems. These oceanic circulation changes highlight how regional climatic events can cascade into broader, global phenomena, altering temperature gradients, precipitation patterns, and ecosystem dynamics.

Impacts on Agriculture, Economies, and Societies

The Little Ice Age had profound and far-reaching effects on agriculture, economies, and societies, as cooler temperatures, shorter growing seasons, and harsher winters created widespread hardship. These impacts were not uniform across regions but were felt globally, with interconnected consequences that shaped historical events and societal responses.

Agriculture and Famines

The Little Ice Age severely affected agricultural productivity due to reduced growing seasons and colder average temperatures. This disruption was particularly pronounced in Europe, where medieval agrarian economies depended heavily on stable climatic conditions. Crops such as wheat, barley, and rye struggled to mature in the cooler conditions, leading to frequent harvest failures. One of the earliest and most devastating examples was the Great Famine of 1315–1317.

Triggered by a series of poor harvests exacerbated by relentless rain and unseasonably cold weather, this famine resulted in the deaths of an estimated 10%–15% of the European population (Campbell, 2016). Food shortages not only led to widespread starvation but also weakened populations, making them more susceptible to disease outbreaks, including the Black Death a few decades later.

The climatic stresses of the Little Ice Age persisted into the 19th century, as seen in the Irish Potato Famine (1845–1852). While primarily caused by a potato blight, the famine was exacerbated by colder and wetter conditions linked to the Little Ice Age, which reduced the resilience of agricultural systems. Millions of Irish people either died or emigrated during this period, fundamentally altering Ireland's demographic and cultural landscape.

Declining agricultural productivity forced societies to adapt by developing new farming techniques and diversifying their food sources. Practices such as crop rotation, selective breeding, and improved irrigation systems gained traction as communities sought to mitigate the risks of future climatic variability. These adaptations laid the groundwork for agricultural innovations in subsequent centuries but came too late to prevent the widespread suffering caused by the Little Ice Age.

Economic Disruptions

The cooling of the Little Ice Age also disrupted economies heavily reliant on agriculture, fishing, and trade. In Northern Europe, expanded sea ice and colder ocean temperatures caused fish stocks to migrate to more temperate waters, undermining local fishing industries. Cod fisheries, critical to the economy of Northern European coastal communities, suffered as fish populations shifted away from traditional fishing grounds.

In Greenland, the Norse settlements, which had thrived during the warmer Medieval Climate Anomaly, struggled to sustain themselves as colder conditions reduced agricultural output and increased sea ice blocked trade routes. By the 15th century, these settlements had collapsed, highlighting the vulnerability of societies dependent on fragile climatic and economic systems.

The broader economic impacts of the Little Ice Age were compounded by social instability. Declining agricultural productivity led to food shortages, which in turn drove inflation and created economic disparities. In urban areas, rising bread prices often sparked riots and social unrest, as seen during the French Flour War of 1775, a precursor to the larger upheavals of the French Revolution.

Societal and Cultural Impacts

The harsh conditions of the Little Ice Age left a significant mark on societal and cultural developments. In Europe, the anxieties and hardships caused by failing crops and harsh winters contributed to an atmosphere of fear and superstition. Scholars have linked the prevalence of **witch hunts** during the 16th and 17th centuries to this period of climatic stress, as societies sought scapegoats for unexplained misfortunes. For example, in Germany, a country severely impacted by the Little Ice Age, accusations of witchcraft often surged after crop failures and harsh winters (Behringer, 1999).

Cold winters, such as the "Great Frost" of 1709, which devastated much of Europe, left lasting impressions on art, literature, and folklore. Artists like Pieter Bruegel the Elder depicted winter scenes in works such as *Hunters in the Snow*, capturing the struggles and beauty of life during the Little Ice Age. Folklore and narratives about the harshness of winter became embedded in the cultural consciousness, influencing storytelling and local traditions.

The Little Ice Age also reshaped societal structures as migration became a survival strategy for many communities. In regions where agricultural systems collapsed, populations moved to areas with more favorable climatic conditions. These migrations often brought cultural exchange but also led to conflict over resources and territory.

Recovery and the Transition to the Modern Warming Period

The recovery from the Little Ice Age and the transition to the modern warming period marked a pivotal moment in Earth's climatic history. Beginning in the mid-19th century, this warming trend coincided with the end of the Dalton Minimum (1790–1830), a period of reduced solar activity, and the dawn of the Industrial Revolution, which introduced significant anthropogenic influences on the climate. The Industrial Revolution catalyzed the widespread use of fossil fuels such as coal, oil, and natural gas, leading to a substantial increase in greenhouse gas emissions and the intensification of the greenhouse effect (IPCC, 2021). These emissions initiated a new era of rapid global warming that continues to shape the modern world.

The retreat of glaciers during this warming period had profound ecological and societal implications. As ice sheets and mountain glaciers receded, ecosystems expanded into newly deglaciated areas, allowing plant and animal species to recolonize and diversify. Forests re-established themselves in higher latitudes and altitudes, while wetlands and grasslands flourished in areas previously dominated by permafrost. This ecological rebound provided renewed resources for human populations, aiding recovery from the economic and social hardships of the Little Ice Age.

Advances in agriculture and industrialization played a crucial role in this recovery. The development of mechanized farming, improved crop varieties, and more efficient irrigation systems increased agricultural productivity, alleviating some of the food security challenges exacerbated by the Little Ice Age. The rise of industrial economies further bolstered societies, providing jobs, infrastructure, and technological innovations. However, these advancements came at a cost: the increased reliance on fossil fuels initiated a new phase of anthropogenic climate change, laying the foundation for the environmental challenges faced today.

The legacy of the Little Ice Age endures in historical records, geological evidence, and societal narratives. It serves as a stark reminder of the profound influence of natural climate variability on human history and the resilience required to adapt to changing environmental conditions. The Little Ice Age also underscores the interconnectedness of Earth's systems, where solar activity, volcanic eruptions, and oceanic circulation patterns interact to produce significant climatic shifts.

As the planet faces unprecedented climate challenges in the modern era, the lessons of the Little Ice Age remain more relevant than ever. This period highlights humanity's vulnerability to climatic fluctuations and the necessity of resilience and adaptation. While the natural forces that shaped the Little Ice Age were beyond human control, today's warming trends are undeniably influenced by anthropogenic activities. Understanding the interplay between natural and human-induced climate drivers provides critical insights into addressing current and future climatic crises. The story of the Little Ice Age reminds us that while we cannot halt all natural climate processes, we have the power and responsibility to mitigate our impact and build a sustainable future for generations to come.

Chapter 12

Earth's Untold Climate Revolutions

Throughout Earth's history, significant climatic events have shaped the planet's ecosystems, geology, and life. These periods, characterized by both abrupt and prolonged shifts in temperature, offer invaluable insights into the dynamics of Earth's climate system and its potential future under anthropogenic influence. This chapter explores the Paleocene-Eocene Thermal Maximum (PETM), the Younger Dryas, and the Medieval Warm Period, highlighting their causes, impacts, and relevance to modern climate change.

Paleocene-Eocene Thermal Maximum (PETM)

A Case Study in Extreme Warming

The Paleocene-Eocene Thermal Maximum (PETM), occurring approximately 56 million years ago, is widely regarded as one of Earth's most dramatic global warming events, offering critical insights into the dynamics of rapid climatic shifts. During the PETM, global temperatures rose by an estimated 5–8°C over a period of just a few thousand years, a rate of warming that profoundly impacted Earth's climate, ecosystems, and carbon cycle. This event was triggered by a massive release of carbon into the atmosphere, likely due to volcanic activity, the destabilization of methane hydrates in ocean sediments, or a combination of these factors (Zachos et al., 2008). The PETM represents a key case study for understanding the interplay between greenhouse gases and global warming.

Geological evidence from marine sediment cores provides compelling records of the PETM's impact on Earth's systems. Elevated atmospheric CO_2 levels during this period caused significant ocean acidification, as the absorption of excess CO_2 lowered the pH of seawater. This acidification was catastrophic for many deep-sea species, particularly those with calcium carbonate shells, which dissolved under the increasingly acidic conditions. The extinction of a large number of deep-sea foraminifera during the PETM is a stark indicator of the vulnerability of marine life to rapid chemical changes in the ocean (Zachos et al., 2008). On land, the warming led to the expansion of tropical and subtropical ecosystems, pushing species to migrate toward higher latitudes in search of suitable habitats.

The terrestrial impacts of the PETM extended beyond shifts in species distributions. Mammalian evolution experienced a significant boost during this time, as changing environments created new ecological niches. Early primates, for example, diversified rapidly, taking advantage of the expanded tropical forests and altered food webs. Fossil evidence suggests that this period marked an important phase in the evolutionary trajectory of mammals, with implications for understanding the adaptive responses of species to environmental pressures (McInerney & Wing, 2011).

The PETM also left a lasting legacy on Earth's carbon cycle. The massive release of carbon altered the balance of the global system, resulting in a prolonged period of elevated temperatures and altered climatic patterns. While the exact source of the carbon release remains debated, possible mechanisms include extensive volcanic activity associated with the North Atlantic Igneous Province and the rapid release of methane from destabilized hydrates in marine sediments

(Dickens, 2011). These events triggered feedback loops, such as the release of additional carbon from terrestrial and oceanic reservoirs, amplifying the warming effect.

One of the most striking aspects of the PETM is its relevance to understanding modern anthropogenic climate change. The rapid increase in greenhouse gases during the PETM mirrors current trends in CO_2 emissions, providing a natural analog for the potential long-term consequences of human-induced climate change. The widespread ecological disruptions and extinctions observed during the PETM highlight the risks of surpassing critical carbon thresholds and the cascading effects that can result from rapid warming (Zachos et al., 2008).

The PETM underscores the fragility of Earth's systems in the face of abrupt climatic shifts and the potential for long-lasting impacts on biodiversity and ecosystems. As we confront unprecedented rates of greenhouse gas emissions today, the lessons of the PETM serve as a cautionary tale about the risks of destabilizing Earth's climate systems and the urgent need for mitigating human impacts.

Younger Dryas: Abrupt Cooling at the End of the Last Ice Age

The Younger Dryas, a sudden and dramatic cooling event that occurred approximately 12,900–11,700 years ago, stands out as one of the most significant climatic shifts during the transition from the last Ice Age to the Holocene Epoch. This period, characterized by an abrupt return to near-glacial conditions, interrupted the prevailing trend of gradual warming. Temperature declines in parts of the Northern Hemisphere were as steep as 10°C within just a few decades, illustrating the rapidity and intensity of this climatic anomaly (Carlson, 2010). Its causes and impacts continue to provide critical insights into the dynamics of Earth's climate system.

The leading hypothesis for the Younger Dryas centers on a disruption of the Atlantic Meridional Overturning Circulation (AMOC), a crucial component of the global thermohaline circulation system. This disruption is believed to have been triggered by a massive influx of freshwater into the North Atlantic, primarily from the melting Laurentide Ice Sheet in North America. The sudden release of this freshwater, possibly through the catastrophic draining of proglacial lakes such as Lake Agassiz, reduced the salinity and density of surface waters, thereby inhibiting the sinking of cold, salty water that drives the AMOC (Broecker, 2006). This slowdown in thermohaline circulation diminished the transport of warm tropical waters to higher latitudes, leading to extensive cooling across much of the Northern Hemisphere.

Evidence of the Younger Dryas is found in various paleoclimatic records, including ice cores, lake sediments, and pollen analyses. Greenland ice cores, for example, reveal a distinct drop in temperatures and a corresponding decrease in greenhouse gas concentrations, such as carbon dioxide and methane, during this period (Alley, 2000). Sedimentary records from lakes in North America and Europe show changes in vegetation patterns, with a decline in tree pollen and an increase in cold-tolerant plant species, indicating a shift to colder and drier conditions.

The Younger Dryas had profound ecological and cultural consequences. In North America, the sudden cooling coincided with the extinction of several megafaunal species, including mammoths, mastodons, and giant ground sloths. While the precise causes of these extinctions remain debated, it is likely that the combined pressures of abrupt climate change and overhunting by early human populations played a critical role (Faith & Surovell, 2009). The loss of these keystone species reshaped ecosystems and forced surviving species to adapt to rapidly changing conditions.

For early human societies, the Younger Dryas presented significant challenges. The abrupt climate shift disrupted food sources, requiring adaptations in subsistence strategies. Archaeological evidence suggests that some human groups adopted more diverse diets, incorporating smaller game, fish, and plant resources to cope with the changing environment. Migration patterns also shifted, as populations moved to areas with more stable climates or abundant resources. These adaptations reflect the resilience and ingenuity of early human societies in the face of environmental upheaval.

The Younger Dryas serves as a stark reminder of the sensitivity of Earth's climate system to disruptions in oceanic circulation. The event underscores the interconnectedness of atmospheric, oceanic, and terrestrial systems and the cascading effects that can arise from even localized changes. As modern climate change accelerates the melting of polar ice caps and glaciers, the potential for similar disruptions to thermohaline circulation raises concerns about the stability of Earth's current climate system. The Younger Dryas provides a valuable historical precedent for understanding the potential consequences of such disruptions, emphasizing the need for proactive measures to mitigate human-induced climate change.

Medieval Warm Period: A Time of Relative Climatic Stability and Warmth

The Medieval Warm Period (MWP), spanning roughly 950–1250 CE, represents a notable period of relative warmth and climatic stability, particularly in parts of Europe, the North Atlantic, and the Arctic. However, the extent and magnitude of the warming were not globally uniform, with significant regional variations. Studies of paleoclimatic proxies, such as tree rings, ice cores, and sediment records, reveal that some regions experienced significant warming while others saw negligible changes or even cooling trends (Mann et al., 2009). This variability underscores the localized nature of the MWP and highlights the complexity of Earth's climate system during this period.

Warming during the MWP is attributed to a combination of natural factors. Increased solar activity, evidenced by a higher number of sunspots, likely contributed to higher energy input into the Earth system. Reduced volcanic activity during this time also decreased the atmospheric aerosols that typically reflect sunlight, further enhancing warming (Eddy, 1976). Additionally, natural variability in oceanic and atmospheric circulation patterns, such as a positive phase of the North Atlantic Oscillation (NAO), likely influenced regional climatic conditions, directing warmer air to northern and western Europe (Trouet et al., 2009).

The MWP had profound social and economic implications, particularly in Europe. Milder temperatures extended growing seasons, allowing agricultural expansion into previously marginal lands. Crops such as barley and wheat thrived, supporting population growth and economic development. Vineyards flourished in regions like southern England, which today are generally considered too cool for widespread viticulture, indicating that the climate was more favorable for certain agricultural activities during this period (Lamb, 1965).

The warming also enabled the Norse to establish and sustain settlements in Greenland. Historical records and archaeological evidence suggest that these settlements benefited from reduced sea ice, which facilitated maritime trade and fishing, as well as milder temperatures that allowed limited farming and livestock grazing (Ogilvie et al., 2000). However, the later cooling associated with the Little Ice Age would eventually contribute to the abandonment of these settlements, illustrating the vulnerability of societies to climatic shifts.

Despite its benefits in some regions, the MWP posed challenges in others. In the American Southwest, for instance, prolonged droughts associated with the MWP led to significant societal stress. Archaeological evidence indicates that the Ancestral Puebloans, who had built complex communities such as those in Chaco Canyon, faced severe water shortages during this period. These conditions likely contributed to the abandonment of these settlements and the migration of populations to areas with more reliable water sources (Benson et al., 2007).

The regional disparities in the impacts of the MWP highlight the complex interplay between climatic shifts and human resilience. While some societies flourished under the warmer conditions, others struggled with droughts, resource scarcity, and changing environmental conditions. These varied responses provide valuable insights into how human societies adapt, or fail to adapt, to climatic changes.

MWP serves as a critical reference point for understanding the natural variability of Earth's climate system. While modern warming trends are largely driven by anthropogenic greenhouse gas emissions, the MWP demonstrates that significant regional warming can occur due to natural factors. Studying this period helps contextualize current climatic changes within the broader framework of historical climate fluctuations, providing a baseline for distinguishing between natural and human-driven climatic shifts. It also underscores the importance of regional studies in assessing the localized impacts of climate variability, which are essential for developing effective adaptation strategies.

How Shifts Inform Our Understanding of Modern Climate Change

Each of these major climate events—the Paleocene-Eocene Thermal Maximum (PETM), Younger Dryas, and Medieval Warm Period (MWP)—offers unique insights into the complexity of Earth's climate system and its profound influence on ecosystems, geology, and human societies. The PETM stands as a stark warning about the catastrophic impacts of rapid greenhouse gas emissions. With a temperature rise of 5–8°C over a few thousand years, the PETM serves as a chilling reminder of how exceeding natural carbon thresholds can destabilize ecosystems, acidify oceans, and trigger biodiversity loss (Zachos et al., 2008). It mirrors the trajectory of modern anthropogenic climate change, emphasizing the long-term consequences of unchecked emissions and the fragility of Earth's carbon cycle.

The Younger Dryas highlights the vulnerability of the climate system to abrupt disruptions, particularly those linked to oceanic circulation. The sudden influx of freshwater into the North Atlantic, disrupting the Atlantic Meridional Overturning Circulation (AMOC), caused rapid cooling and dramatic ecological and cultural shifts (Broecker, 2006). This event underscores the importance of maintaining the delicate balance of Earth's thermohaline circulation, a system increasingly at risk from polar ice melt in the modern era.

The MWP, though less dramatic in magnitude, underscores the influence of regional climatic variations on human societies, resource availability, and ecosystems. In Europe, it facilitated agricultural expansion and population growth, while in other regions, such as the American Southwest, it brought prolonged droughts that challenged human resilience (Mann et al., 2009). The MWP exemplifies how even relatively moderate climatic shifts can have widely varying impacts across regions, highlighting the importance of studying localized effects within the context of broader climate trends.

Collectively, these historical climate events emphasize the intricate interplay between natural variability and external forces, both natural and anthropogenic. They illustrate how changes in one component of Earth's climate system, whether greenhouse gas levels, ocean circulation, or solar activity, can cascade through interconnected systems, triggering widespread transformations.

Understanding past climate events is not merely an academic exercise; it is a roadmap for navigating the challenges of today's rapidly warming world. The PETM warns of the long-term ecological and geological consequences of elevated greenhouse gas emissions. The Younger Dryas demonstrates the potential for abrupt and unpredictable shifts when critical thresholds are breached. The MWP, with its diverse regional impacts, reveals how societies respond to climatic opportunities and challenges. These events collectively underscore that Earth's climate system is both resilient and profoundly sensitive, capable of recovering over millennia but also vulnerable to catastrophic shifts.

As we face unprecedented levels of anthropogenic warming, these lessons from history remind us of the urgency to mitigate greenhouse gas emissions, safeguard critical Earth systems, and adapt to inevitable changes. They also highlight the interconnectedness of human actions and natural processes, demonstrating that the choices we make today will shape the trajectory of Earth's climate and ecosystems for centuries to come. By heeding the lessons of the past, we can chart a course toward a more sustainable and resilient future.

Chapter 13

Civilizations Lost to Climate Change

History is littered with the ruins of civilizations that once thrived, mighty empires reduced to stone remnants and whispered legends. While war and conquest often take the blame, a quieter force has been the true architect of collapse time and again: climate. Shifts in rainfall, prolonged droughts, sudden cold snaps, these changes in Earth's rhythm have brought societies to their knees long before carbon emissions were a concern. From the cliff dwellings of the Ancestral Puebloans to the rainforest cities of the Maya, the story is alarmingly familiar: environmental stress escalates, systems unravel, and the center cannot hold. In this chapter, we journey through the archaeological and climatological records to uncover how climate change, natural, often abrupt, and merciless, has toppled some of history's most sophisticated societies. Their stories serve as ancient cautionary tales for our modern age, reminding us that nature always has the final word.

Echoes in the Dust

Climate change isn't just a modern threat, it's an ancient executioner. Long before humanity burned its first barrel of oil, entire civilizations had already risen and vanished, not in the fury of war or under the weight of economic collapse, but beneath the slow, unrelenting pressure of a shifting climate. These ancient societies, once thriving and sophisticated, were undone not by external invaders but by internal vulnerabilities exposed by environmental instability. As rainfall patterns changed, rivers dried, and fertile lands turned barren, complex cultures crumbled into silence. This chapter explores four such civilizations, the Ancestral Puebloans, the Maya, the Akkadian Empire, and the Norse settlers of Greenland, whose dramatic declines have been linked to abrupt and prolonged climatic shifts. Their stories are not relics of a distant past; they are warnings etched in stone and soil, sounding alarms for a world now entering its own era of instability.

The Ancestral Puebloans (Anasazi) – Drought in the Desert

The Ancestral Puebloans, often referred to by the now-retired term "Anasazi," developed one of the most remarkable prehistoric cultures in North America. Flourishing from approximately 900 to 1300 CE, they built monumental structures and complex communities in what is now the Four Corners region of the American Southwest, including the iconic sites of Chaco Canyon and Mesa Verde. These communities featured multi-storied stone dwellings, ceremonial kivas, and sophisticated irrigation systems, testaments to their advanced understanding of engineering and environmental management (Lekson, 2006).

Despite their ingenuity, the Ancestral Puebloans were not immune to climate. By the late 1200s, many of their settlements, particularly the larger ones, were abruptly abandoned. Archaeological and dendroclimatological studies have since revealed that a prolonged and severe drought, lasting from roughly 1275 to 1300 CE, coincided with this widespread abandonment (Dean & Funkhouser, 2002). Tree ring records, one of the most reliable paleoclimate indicators, show an extended period of minimal precipitation across the Colorado Plateau, a region already marked by arid conditions.

This so-called "Great Drought" had devastating effects. As water supplies dwindled and crop yields declined, once-cooperative communities may have turned inward and competitive. Evidence suggests increased fortification of settlements and a rise in violent conflict during this period (Kohler et al., 2008). Faced with failing agricultural systems and resource scarcity, populations began to migrate, likely integrating into other Puebloan groups farther east and south, such as the ancestors of today's Hopi and Zuni peoples.

What makes the story of the Ancestral Puebloans so resonant today is not just their collapse, but the nature of it. They were a society that pushed the environmental limits of their landscape, building large populations in a semi-arid region reliant on fragile agricultural systems and limited water sources. When climate pushed that system beyond its tipping point, the collapse was swift and widespread. Their legacy is both inspiring and cautionary: even the most advanced societies can falter when they exceed the carrying capacity of their environment.

The Maya – Rainforest Empire with Dry Bones

The Maya civilization, known for its awe-inspiring pyramids, intricate glyphs, and complex calendrical systems, thrived across much of Mesoamerica from roughly 2000 BCE to the 1500s CE. At its height during the Classic Period (c. 250–900 CE), the southern lowland region of modern-day Guatemala, Belize, and parts of Mexico was dotted with bustling Maya city-states like Tikal, Copán, and Palenque. These urban centers were political, economic, and ceremonial hubs, connected by trade routes and governed by a class of elites who claimed divine descent (Martin & Grube, 2008).

Yet by the 9th century CE, many of these monumental cities were mysteriously abandoned. The collapse was not uniform or instantaneous, but it was widespread and dramatic. Population centers emptied, monumental architecture ceased, and the centralized political structures unraveled. For decades, scholars debated the causes of this

"Classic Maya collapse." Today, a growing body of multidisciplinary evidence points to one inescapable culprit: climate change, specifically, a series of prolonged and intensifying droughts, compounded by environmental mismanagement (Douglas et al., 2015).

High-resolution climate data extracted from lake sediment cores and speleothems (cave stalagmites) paint a sobering picture. A major study of a stalagmite from Yok Balum Cave in Belize revealed that between 800 and 1000 CE, the region experienced several severe, multi-decade droughts, including an especially intense dry spell beginning around 820 CE that coincided with the collapse of several major Maya cities (Kennett et al., 2012). Similarly, sediment core analyses from nearby lakes show lower water levels, reduced vegetation cover, and increased indicators of soil erosion, clear signs of drought and deforestation (Hodell et al., 1995).

The Maya, despite their advanced knowledge of astronomy and engineering, had pushed their environment to the brink. They practiced slash-and-burn agriculture, cleared vast swaths of rainforest for urban expansion, and constructed large reservoirs and canals that were ultimately vulnerable to prolonged dry conditions. The overuse of these fragile systems during periods of climatic stress led to agricultural failure, food shortages, and social upheaval. As political legitimacy eroded, elite structures collapsed, and many cities were simply abandoned (Lucero, 2002).

This unraveling of complexity is a stark reminder of what can happen when a civilization outpaces its ecological limits. The Maya's fall wasn't due to a single cataclysm, but to the slow, compounding failure of systems unable to adapt to a drying world. The lesson is chilling in its modern resonance: even a society as intellectually sophisticated as the Maya can be brought down when nature's balance is broken and the human response is inflexible.

The Akkadian Empire – Dust and Decline

More than four thousand years ago, the Akkadian Empire emerged in Mesopotamia as the world's first true empire, uniting the Sumerian city-states under a centralized rule led by Sargon of Akkad around 2334 BCE. At its height, the empire stretched from the Persian Gulf to the Mediterranean, a marvel of administrative control, military innovation, and urban development. It set the template for imperial governance, establishing a model that would echo throughout history (Weiss et al., 1993).

And yet, despite its might, the Akkadian Empire collapsed with alarming speed around 2200 BCE. For a long time, historians attributed its fall to political instability or internal rebellion. But compelling geological and archaeological evidence has reshaped this narrative: the empire's sudden disintegration coincided with a catastrophic, long-term climate event known as the 4.2 kiloyear event, a roughly 300-year megadrought that devastated the Fertile Crescent (Weiss & Bradley, 2001).

Ocean sediment cores from the Gulf of Oman, along with dust deposition records in Mesopotamia, suggest a sharp reduction in monsoon rainfall around 2200 BCE. This decline in precipitation triggered widespread aridification of the region, drastically reducing crop yields and drying out once-fertile lands. Soil records and paleoclimate indicators from Syria and northern Iraq point to a sudden increase in wind-blown dust, indicative of desertification and the collapse of agriculture (Cullen et al., 2000). The formerly lush, grain-producing heartland of Akkad turned to dust.

As food supplies dwindled, mass migrations from northern regions into the southern urban centers followed, overwhelming local systems and triggering violent conflict. Archaeological records reveal abandoned settlements, fortified towns, and signs of population displacement during this period (Weiss et al., 1993). The famed city of

Akkad itself, once the seat of imperial power, vanished from history, its exact location still unknown.

This collapse underscores a sobering truth: even the most advanced political systems and military structures are powerless in the face of sustained environmental breakdown. The Akkadians had developed irrigation networks, trade systems, and monumental infrastructure, but none of it could outlast centuries of drought. Their fall is a testament to how ecological fragility can unmake civilizations built on the illusion of control.

Norse Greenland – Frozen Out by the Little Ice Age

In the late 10th century, Norse explorers led by Erik the Red established settlements along the southwestern coast of Greenland. These hardy communities, known as the Eastern and Western Settlements, were outposts of Norse civilization at the edge of the known world. For nearly 400 years, Norse Greenlanders built churches, raised livestock, and traded with Europe, sustaining themselves in an unforgiving environment through sheer determination and rigid adherence to familiar ways of life (Diamond, 2005).

But by the early 15th century, these settlements had vanished. The churches fell silent, the pastures overgrew, and the last written records, including a wedding in 1408, gave way to silence. Archaeological evidence and historical climatology now point to a major environmental cause for this disappearance: the onset of the Little Ice Age, a period of global cooling that began around 1300 CE and lasted into the 19th century (Ogilvie & Jónsson, 2001). For Greenland's Norse colonies, this cooling came with devastating consequences.

Ice core data and paleotemperature reconstructions indicate that Greenland experienced a drop in average temperatures by several degrees during the 14th and 15th centuries (Dugmore et al., 2007). This cooling shortened growing seasons and expanded sea ice, making it increasingly difficult for Norse farmers to grow hay or maintain livestock, both essential to their dairy-based economy. The icy waters also made transatlantic trade routes impassable, severing connections with Norway and Iceland, and leaving the settlers dangerously isolated.

Famine and malnutrition followed. Isotope analysis of Norse Greenlandic bones shows a shift from a land-based diet to one increasingly reliant on marine food sources as crops failed and herds declined (Arneborg et al., 1999). Yet even as conditions worsened, the Norse refused to fully adapt. They maintained their European-style agriculture and clothing, and crucially, failed to adopt the more sustainable subsistence strategies of their Inuit neighbors, such as seal hunting, skin-based clothing, and mobile shelter (McGovern, 1990).

This cultural rigidity proved fatal. Unlike the Inuit, who thrived during the same climatic downturn, the Norse Greenlanders disappeared— whether through starvation, outmigration, or integration remains debated, but their permanent settlements did not survive. Their story offers one of the clearest historical examples of how environmental inflexibility in the face of climate change can lead to societal collapse.

In the saga of Norse Greenland, the lesson is stark: adaptability is survival. Their fate underscores the dangers of clinging to tradition when the world changes around you, and the price of ignoring those who have already learned to thrive in difficult environments.

Patterns of Collapse – What They All Share

Across time and continents, the fall of the Ancestral Puebloans, the Maya, the Akkadian Empire, and the Norse Greenlanders may seem like disconnected events, products of local geography, culture, and circumstance. But when viewed through the lens of climate history, they form a chillingly coherent pattern. Each of these civilizations faced environmental changes that tested the limits of their resilience, and each ultimately failed to adapt in time. Their stories echo across centuries with unsettling familiarity, revealing three interconnected threads that bind their downfalls.

Environmental Stress plus Poor Adaptation

At the heart of every collapse was a changing climate. Not a single, cataclysmic event, but a prolonged shift, a creeping drought, a steady cooling, a drying wind, that undermined the very systems upon which these societies depended (Weiss & Bradley, 2001; Kennett et al., 2012; Dugmore et al., 2007). But climate alone didn't doom them. It was the inability, or refusal, to adapt that sealed their fate. The Maya continued deforestation during drought; the Akkadians stayed as their fields turned to dust; the Norse Greenlanders clung to their cattle even as snow buried their pastures. Nature posed the challenge, but it was human inflexibility that triggered the fall (Diamond, 2005).

Overreliance on Fragile Systems

Each society built its success on systems finely tuned to a specific climate regime, irrigation networks in Mesopotamia, maize farming in the Southwest, maritime trade in the North Atlantic. These systems were remarkably efficient when conditions were stable. But they were also brittle (Cullen et al., 2000; Lucero, 2002). A few degrees of cooling, a decade without rain, or a shift in ocean currents was enough to tip them into failure. The more specialized and interconnected these systems became, the more vulnerable they were to disruption. Complexity became a liability when resilience was what was needed most.

Sociopolitical Fragility in the Face of Ecological Disruption

As environmental stress intensified, so too did social fragmentation. Political elites lost legitimacy. Trade routes collapsed. Civil unrest and migration rose. Whether it was Maya rulers who could no longer summon rain through ritual, Akkadian kings who could no longer feed their armies, or Puebloan leaders unable to prevent famine, authority faltered (Kohler et al., 2008; Weiss et al., 1993). The social glue that had once bound their civilizations together dissolved under the pressure of unmet needs and growing inequality. In all four cases, ecological collapse triggered a societal unravelling from the top down.

These are not just stories of the past. They are warnings for the present.

Could It Happen to Us?

It's tempting to see these ancient collapses as the unfortunate consequences of ignorance or technological limits. But what if they're not that different from us?

Today, cities like Los Angeles, Cape Town, and Chennai have already come dangerously close to "Day Zero" water crises (Marlow et al., 2021). The American West faces historic megadroughts, with reservoirs at record lows and agriculture under strain (Williams et al., 2020). Meanwhile, rising seas threaten coastal megacities from Miami to Jakarta, and global food supply chains, once seen as marvels of modern efficiency, have shown just how fragile they really are.

Like the Maya, we're clearing forests faster than they can recover. Like the Akkadians, we depend on monoculture crops and rivers whose flows are increasingly erratic. Like the Norse, we often resist change, clinging to traditions and comforts that may no longer be sustainable. And like the Puebloans, we too face choices about migration, adaptation, and survival under climatic stress.

The difference? We have the knowledge, and the warning signs. But will we act?

The fall of ancient civilizations is not just a history lesson; it's a mirror. The patterns are clear. The question is whether we will break the cycle or become just another story etched into the dust of time.

Tomorrow's Archaeologists

Centuries from now, long after our voices have faded, what will tomorrow's archaeologists find buried in the bones of our civilization?

Will they unearth plastic-entombed coastlines where coral reefs once thrived? Will they sift through the concrete skeletons of drowned cities and wonder how we didn't see it coming, or worse, how we saw it and still did nothing? Will they marvel at our brilliance… and shake their heads at our blindness?

Every collapsed civilization in this chapter left behind a signature, traces of a society that reached too far, too fast, and couldn't bend before it broke. The Ancestral Puebloans, the Maya, the Akkadians, the Norse, all victims not just of climate change, but of climate indifference, of systems built on narrow margins and traditions too sacred to abandon. Their fall wasn't sudden. It was slow erosion masked by the illusion of stability, until it wasn't.

And now, the cycle has come full circle. But this time, we are the climate shift. This time, we are not waiting on a megadrought, or a cooler sun. Our industries, our lifestyles, our choices have become geological forces, carving the future in carbon and methane. We are rewriting the planet's operating system, and we're doing it faster than nature has ever done before.

The question is no longer if we will be remembered, but how. Will we be the civilization that finally broke the cycle of collapse, or just another cautionary tale written in soot and sediment?

As we turn the page, we examine what may be the most consequential chapter of all: how humanity's accelerating footprint is reshaping Earth's climate today. Not as passengers, but as drivers of the storm.

Chapter 14

Lessons from the Ice

Earth's climate history offers a compelling narrative of transformation and resilience, providing critical insights into how the planet's systems respond to natural and anthropogenic changes. From the dramatic shifts of glaciations and interglacials to extreme events like the Paleocene-Eocene Thermal Maximum (PETM) and the Younger Dryas, these periods underscore the intricate connections between climate, ecosystems, and human civilizations. This chapter synthesizes these lessons, explores their implications for future climate resilience, and examines the role of science, technology, and policy in navigating the challenges of a rapidly changing climate.

Synthesizing the Lessons from Earth's Climate History

Earth's climate history offers a profound narrative of resilience and sensitivity, demonstrating how the planet's systems respond to both gradual and abrupt changes. Glaciations, defined by the advance of massive ice sheets, have repeatedly reshaped landscapes and ecosystems, carving valleys, redirecting rivers, and depositing sediments that formed the basis for fertile soils. These periods of extensive glaciation forced species to adapt to harsh conditions, migrate to more hospitable environments, or face extinction. For instance, megafaunal species such as mammoths and saber-toothed cats thrived during glacial periods but ultimately succumbed to environmental changes and human pressures at the end of the last Ice Age (Faith & Surovell, 2009). Glacial cycles acted as evolutionary crucibles, driving the emergence of resilient species and ecosystems.

Conversely, interglacial periods marked by warmer temperatures provided the stability needed for ecosystems to recover and diversify. Forests, grasslands, and wetlands expanded into deglaciated regions, fostering biodiversity and creating niches for new species. These warmer intervals also provided opportunities for human civilizations to emerge and flourish. The Holocene, the current interglacial period, has supported agriculture, trade, and societal complexity by offering a relatively stable climate that allowed human populations to grow and innovate (Ponting, 2007). These cycles highlight the interconnectedness of Earth's systems, where changes in one component, whether in ocean circulation, atmospheric composition, or solar radiation, trigger cascading effects that shape global climate and ecosystems.

Historical climate events, such as the Paleocene-Eocene Thermal Maximum (PETM), underscore the catastrophic potential of rapid greenhouse gas emissions. During the PETM, global temperatures increased by 5–8°C due to a massive release of carbon into the atmosphere, likely from volcanic activity or the destabilization of methane hydrates (Zachos et al., 2008). The event triggered widespread ocean acidification, biodiversity loss, and altered weather patterns, demonstrating the long-term impacts of exceeding carbon thresholds. These lessons resonate today as modern anthropogenic emissions accelerate at an unprecedented rate, threatening to replicate similar outcomes on a global scale.

The Younger Dryas, an abrupt cooling event approximately 12,900 years ago, underscores the sensitivity of Earth's climate to rapid shifts. Triggered by the disruption of the Atlantic Meridional Overturning Circulation (AMOC) due to a sudden influx of freshwater from melting ice sheets, this event led to a dramatic temperature drop of up to 10°C in parts of the Northern Hemisphere (Broecker, 2006). The Younger Dryas serves as a stark reminder of how critical thresholds in oceanic circulation can cascade into broader climatic changes, disrupting ecosystems and human societies alike. As modern polar ice melt accelerates, the lessons of the Younger Dryas are particularly pertinent for understanding the risks of current and future disruptions to the AMOC.

These historical events not only illuminate the resilience of Earth's systems but also reveal their profound vulnerability to disruptions. They provide analogs for understanding the mechanisms driving modern climate change, offering both cautionary tales and guidance for action. The PETM, Younger Dryas, and other climatic episodes emphasize the interconnected nature of Earth's systems, demonstrating that changes in atmospheric greenhouse gases, ocean circulation, or land use can reverberate across the globe. As human activities continue to alter these systems, the insights gleaned from Earth's climate history are invaluable for navigating the challenges of the Anthropocene.

Glaciations and Interglacials Changed Human Civilizations

Climate changes have had a transformative impact on ecosystems, shaping both the physical landscape and the course of evolutionary history. The expansion and retreat of ice sheets during the Quaternary

Period, especially during the Pleistocene, sculpted vast swaths of the Earth's surface, creating new landforms and altering existing ones. The movement of glaciers carved out valleys, fjords, and mountain ranges, which remain some of the most recognizable features of Earth's landscape today. Additionally, the advance and retreat of ice sheets helped form fertile soils in regions such as the Great Plains of North America and the Baltic lowlands of Europe. These areas, enriched by glacial till, became critical agricultural hubs, supporting the development of human societies that relied on farming to sustain growing populations (Ehlers & Gibbard, 2007). The soils left behind by glaciations were often more nutrient-rich than those found in other areas, making these regions ideal for the cultivation of a wide variety of crops.

Glacial meltwater played a significant role in shaping the hydrological systems of the planet. As glaciers retreated, they deposited vast amounts of sediment, creating rivers, lakes, and wetlands that would go on to support diverse ecosystems. In North America, for instance, the Great Lakes were formed by the meltwater of retreating glaciers, and today, they represent the largest freshwater system in the world, providing vital resources for biodiversity and human societies. Similarly, the Baltic Sea formed in Europe as the ice sheets retreated, leaving behind a body of water that serves as a key ecological and economic resource for the surrounding regions. Glacial lakes and rivers often provided fertile grounds for the development of human settlements, fostering trade, agriculture, and urbanization in the areas they created (Ehlers & Gibbard, 2007).

Interglacial periods, such as the Holocene, have played an even more significant role in the development of human civilizations. The warm, stable climate of the Holocene, which began approximately 11,700 years ago, has allowed humans to develop agriculture, establish complex societies, and build trade networks that span the globe. The more temperate conditions during this period have enabled the expansion of ecosystems that were once inhospitable during glacial periods. Forests, grasslands, and wetlands spread into regions that had been covered by ice, providing diverse habitats for flora and fauna. The availability of resources in these newly habitable regions fostered the growth of early agricultural societies, which could now settle in one place rather than depending on nomadic hunting and gathering. The development of civilizations in regions such as Mesopotamia, Egypt, and the Indus Valley demonstrates the relationship between climate stability and societal growth, where a predictable and stable climate allowed humans to build cities, engage in trade, and create sophisticated social structures (Ponting, 2007).

However, the relationship between climate and human activity has not always been one of benefit. The Medieval Warm Period (MWP), which spanned roughly 950–1250 CE, is an example of a climatic shift that brought both advantages and challenges. In Europe, the warming allowed for agricultural expansion, increased food production, and population growth. However, in other parts of the world, the impacts of this warming were not as beneficial. For instance, in the American Southwest, the prolonged droughts associated with the MWP severely stressed the agricultural practices of the Ancestral Puebloans. These droughts, which coincided with the warming period, are thought to have contributed to the collapse of their societies, particularly those in the Colorado Plateau and surrounding regions (Benson et al., 2007). This example highlights the dual nature of climate impacts: while warmer conditions may support agricultural growth and prosperity in some regions, they can also lead to resource scarcity, water shortages, and societal collapse in others. The experience of the Ancestral Puebloans illustrates the vulnerability of human societies to shifts in climate, even when those shifts appear to be beneficial in other areas.

The interplay between climate and human activity underscores the importance of adaptability in the face of changing conditions. While interglacial periods, particularly the Holocene, have allowed human civilizations to flourish, the challenges posed by climate shifts— whether from natural causes like the Medieval Warm Period or human-induced factors such as modern global warming—demonstrate the need for strategies that help societies adapt to new climatic realities. The ability to respond to environmental challenges, be it through technological innovation, resource management, or social cooperation, is key to ensuring that human societies can thrive in an ever-changing climate.

Implications for Future Climate Resilience and Adaptation

As the planet faces unprecedented rates of warming, the lessons from Earth's climate history are more relevant than ever. Modern climate change, driven primarily by anthropogenic greenhouse gas emissions, poses risks that mirror those of historical climate events, but occur at an accelerated pace. The Paleocene-Eocene Thermal Maximum (PETM), for example, serves as a stark warning of the long-term impacts of unchecked carbon emissions. During the PETM, the rapid release of carbon into the atmosphere resulted in a dramatic global temperature rise of 5–8°C over a few thousand years. This warming event triggered ocean acidification, a significant biodiversity loss, and altered weather patterns in ways that closely resemble the consequences of today's rising carbon concentrations (McInerney & Wing, 2011). Modern rates of carbon emissions are occurring far more rapidly than during the PETM, raising concerns about similar, or even more severe, long-term ecological and climatic disruptions. As we continue to increase the levels of carbon dioxide in the atmosphere, the risks of ocean acidification, loss of marine and terrestrial species, and disruptions to agricultural systems become ever more pressing.

Similarly, the Younger Dryas, a sudden cooling event approximately 12,900 years ago, highlights how climate systems can undergo abrupt shifts in response to changes in ocean circulation. The Younger Dryas was triggered by a massive influx of freshwater into the North Atlantic, which disrupted the Atlantic Meridional Overturning Circulation (AMOC), a key component of Earth's climate system that helps regulate temperatures in the Northern Hemisphere (Rahmstorf et al., 2005). This disruption caused a rapid return to near-glacial conditions, dramatically affecting ecosystems and human societies. The ongoing melting of polar ice caps, particularly in Greenland, has the potential to destabilize the AMOC, raising concerns about the possibility of abrupt climate shifts in the future. Understanding how ocean circulation has shifted in the past provides crucial insights into the potential impacts of modern changes to the Arctic and Antarctic regions, which may trigger disruptions similar to the Younger Dryas.

Building climate resilience in the face of these risks requires a deep understanding of the thresholds and feedback mechanisms that govern Earth's systems. Our current understanding of climate dynamics—particularly how various systems, such as atmospheric composition, ocean currents, and land surfaces, interact—underscores the fragility of the climate system and the urgency of taking action. One key strategy for mitigating climate change is to protect and restore critical ecosystems that act as carbon sinks, such as forests, wetlands, and peatlands. These ecosystems not only absorb carbon dioxide from the atmosphere but also provide a range of other vital ecological services, including water filtration, habitat for wildlife, and flood regulation. The destruction of these ecosystems through deforestation, urbanization, and industrial activities is a major driver of greenhouse gas emissions, exacerbating climate change. Protecting these natural resources and investing in restoration efforts can significantly enhance the planet's ability to sequester carbon and help mitigate the impacts of global warming.

In addition to ecosystem protection, investments in sustainable agriculture, water management, and renewable energy are essential for reducing vulnerabilities to climate impacts while fostering long-term resilience. Sustainable agricultural practices, such as regenerative farming, agroforestry, and improved water use, can help ensure food security while reducing emissions and enhancing soil health. Water management strategies, including the restoration of wetlands and the use of water-efficient technologies, are crucial for addressing water scarcity and reducing the risk of floods and droughts. Transitioning to renewable energy sources, such as solar, wind, and hydropower, is one of the most effective ways to reduce global carbon emissions, as the burning of fossil fuels for energy production is a major contributor to climate change.

The challenges posed by modern climate change are immense, but they are not insurmountable. By learning from past climate events, investing in sustainable technologies, and implementing policies that protect both the environment and human societies, we can mitigate the risks of future climate disruptions and build a more resilient and sustainable world for future generations.

Science, Technology, and Policy in Mitigating Modern Challenges

Science and technology are central to tackling the pressing climate challenges of our time. Advances in climate modeling, for example, have vastly improved our ability to predict future climate trends and their potential impacts on ecosystems, societies, and economies. Sophisticated computer models simulate a wide range of scenarios, taking into account variables such as greenhouse gas emissions, solar radiation, ocean currents, and land-use changes. These models allow scientists to project temperature rise, sea-level changes, and extreme weather events, providing policymakers with the necessary information to make informed decisions (IPCC, 2021). By integrating real-time data from satellite monitoring systems, scientists can track critical phenomena like deforestation, glacial retreat, and sea-level rise, which are essential indicators of climate change. These technologies not only improve our understanding of the Earth's changing systems but also equip decision-makers with timely data to guide policy, aid in disaster preparedness, and inform mitigation strategies.

Renewable energy innovations are also key in reducing the carbon footprint of human activities. The development of solar, wind, hydroelectric, and geothermal power has the potential to replace fossil fuels as primary energy sources, dramatically lowering carbon emissions. Solar and wind energy, in particular, have seen significant advances in efficiency and cost reduction, making them increasingly competitive with traditional energy sources. For instance, the cost of solar photovoltaic panels has fallen by over 80% since 2010, making them more accessible for both individuals and large-scale energy producers (International Renewable Energy Agency [IRENA], 2020). In addition, breakthroughs in battery storage technology are helping to address the intermittent nature of renewable energy, ensuring a more stable and reliable energy grid. The integration of renewable energy into the global power grid is one of the most effective strategies for mitigating greenhouse gas emissions and transitioning toward a low-carbon economy.

Another promising technological innovation is carbon capture and storage (CCS). CCS technologies are designed to capture carbon dioxide (CO_2) emissions from power plants and industrial processes before they enter the atmosphere and store them underground or in other safe locations. While the widespread implementation of CCS remains challenging, ongoing research is improving its efficiency and scalability. Additionally, direct air capture (DAC) technologies are being developed to remove CO_2 directly from the atmosphere, offering a potential solution for addressing historical emissions and achieving net-zero targets. These technologies, though still in the early stages of deployment, could play a critical role in reducing atmospheric CO_2 concentrations and mitigating climate change impacts.

Sustainable infrastructure is another key component of climate change mitigation. Green buildings, energy-efficient transportation systems, and smart grids reduce emissions and enhance the sustainability of urban environments. Cities, which are responsible for a significant portion of global emissions, can play a central role in addressing climate change by implementing energy-efficient construction practices, promoting public transportation, and integrating green spaces. Moreover, climate-resilient infrastructure that can withstand extreme weather events, such as flood-resistant buildings and renewable-powered water management systems, will be essential as climate change increases the frequency and intensity of such events.

Policy frameworks are equally essential in driving meaningful global climate action. International agreements such as the Paris Agreement, which aims to limit global warming to 1.5°C above pre-industrial levels, have been instrumental in setting global climate targets and fostering international cooperation. However, these agreements alone are insufficient without strong national and local policy actions. National governments must implement policies that incentivize clean energy development, promote reforestation, and adopt climate adaptation strategies. Carbon pricing mechanisms, such as carbon taxes and cap-and-trade systems, can incentivize businesses to reduce emissions by assigning a cost to carbon pollution. Reforestation and afforestation efforts are also critical for sequestering carbon and restoring degraded ecosystems. Policies that focus on energy transition, such as subsidies for renewable energy and investments in green technology, can create jobs, stimulate economic growth, and accelerate the shift away from fossil fuels.

Public awareness and education are crucial in the fight against climate change. Engaging communities in climate action fosters a sense of shared responsibility and empowers individuals to contribute to sustainability efforts. Education on climate science, sustainable practices, and the urgency of action can motivate individuals, businesses, and governments to adopt more sustainable lifestyles and policies. Social movements and grassroots initiatives have proven effective in driving change, as seen in the global rise of environmental activism led by organizations like Greenpeace and movements such as Fridays for Future. Increasingly, consumers are demanding climate-conscious products, and investors are prioritizing companies that prioritize sustainability. Public pressure is crucial for holding governments and corporations accountable for their role in mitigating climate change.

By learning from past climatic events and applying these lessons to contemporary challenges, humanity can navigate the complexities of climate change more effectively. Events such as the Paleocene-Eocene Thermal Maximum (PETM) and the Younger Dryas provide important analogs for understanding the potential consequences of modern climate disruptions. By leveraging science, technology, and policy, societies can build resilience to climate impacts while simultaneously reducing emissions and working toward a sustainable future. Climate change is not an insurmountable challenge; with concerted action, the planet can transition toward a more sustainable, equitable, and resilient future for all.

Earth's climate history is a testament to the resilience and adaptability of natural and human systems in the face of profound changes. The lessons of glaciations, interglacials, and extreme events like the PETM and Younger Dryas reveal the interconnectedness of Earth's systems and the cascading impacts of climatic shifts. These historical events underscore the urgency of addressing modern climate challenges, where the stakes are higher and the pace of change is unprecedented.

As humanity confronts the dual imperatives of mitigation and adaptation, the insights from Earth's climate past provide a roadmap for action. By leveraging scientific understanding, technological innovation, and robust policy frameworks, we can mitigate the impacts of climate change and build a future that respects the delicate balance of Earth's systems. The story of Earth's climate is not only one of transformation but also one of resilience and hope—a reminder that, with collective effort, the challenges of today can become the opportunities of tomorrow.

Chapter 15

Humanity Accelerating the Warm-Up

The Earth's climate is shaped by a variety of natural and anthropogenic (human-driven) factors. Historically, natural drivers such as volcanic activity, solar radiation, and Earth's axial tilt have influenced the climate over geological timescales. However, the recent acceleration of global warming presents a stark departure from these long-term natural fluctuations. Since the Industrial Revolution, human activities have emerged as the dominant force driving climate change, with carbon dioxide (CO_2), methane (CH_4), and other greenhouse gases increasingly filling the atmosphere at unprecedented rates (Intergovernmental Panel on Climate Change [IPCC], 2021). This shift, from primarily natural to anthropogenic drivers, is now evident in the rising global temperatures and increasingly extreme weather events. Urgency is required in addressing the role of human influence, as current trajectories suggest that without significant changes, the planet may reach critical tipping points that will lead to irreversible damage (Rockström et al., 2009).

The Industrial Revolution and Beyond

The Industrial Revolution, beginning in the late 18th century, marked the onset of rapid industrialization, urbanization, and a fundamental transformation in energy production. This period saw the introduction of mechanized manufacturing, shifting economies from agrarian-based societies to industrial powerhouses. Central to this transformation was the widespread use of coal, followed by oil and natural gas, which powered the expansion of industry, transportation, and electricity generation. The reliance on these fossil fuels allowed for unprecedented growth in productivity and living standards, particularly in Western Europe and North America (Jones & Durbin, 2018). However, this rapid expansion also led to a significant alteration in the balance of greenhouse gases in the atmosphere. Fossil fuels, rich in carbon, release large quantities of CO_2 when burned for energy, which has become a key driver of global warming (IPCC, 2021). The combustion of coal, followed by oil and natural gas, released significant amounts of CO_2 into the atmosphere, resulting in the first substantial increase in CO_2 concentrations from pre-industrial levels of approximately 280 parts per million (ppm) to over 400 ppm today (Lindsey, 2021).

This dramatic increase in greenhouse gases coincided with a period of immense population growth and urban expansion. The 19th and 20th centuries saw the global population surge, coupled with the growth of cities as centers of industrial activity. The urbanization of society, driven by industrialization, intensified energy demands, contributing to higher fossil fuel consumption and increased emissions (IPCC, 2021). Furthermore, urban areas became hubs for industrial activity, infrastructure development, and transportation systems that relied heavily on fossil fuels. These urban expansions also led to significant land-use changes, as cities grew outward and agricultural practices intensified to meet the demands of expanding populations. In this context, the planet's energy balance shifted decisively toward anthropogenic causes of climate change, with human activity now driving the unprecedented rise in greenhouse gas emissions (IPCC, 2021). The transformative nature of the Industrial Revolution and its long-lasting impacts on climate underscore the profound influence of industrial activity on Earth's climate systems.

Key Human Activities Driving Climate Change

Fossil Fuel Combustion

The burning of fossil fuels for energy production, transportation, and industry remains the primary source of greenhouse gas emissions today. Fossil fuels—coal, oil, and natural gas—account for approximately 75% of global CO_2 emissions, a staggering proportion that underscores their central role in driving climate change (International Energy Agency [IEA], 2020). The combustion of these fuels occurs across multiple sectors, including electricity generation, transportation, and industrial operations such as manufacturing and chemical production. Power plants, for example, burn coal, oil, or natural gas to produce electricity, while transportation systems rely on petroleum products, such as gasoline and diesel, to fuel vehicles and ships. Additionally, industrial sectors like cement manufacturing and chemical production also contribute significantly to CO_2 emissions, particularly through processes that involve the burning of fossil fuels for heat and energy (IEA, 2020).

The global transportation sector alone is responsible for nearly 14% of total greenhouse gas emissions, with road transport accounting for the majority of these emissions through the combustion of petroleum products (IEA, 2020). The rising demand for vehicles and freight transportation, particularly in rapidly industrializing nations, continues to drive emissions higher. The increased use of air travel and the expansion of global trade further contribute to these emissions, placing additional strain on the planet's climate system. In recent decades, CO_2 concentrations have surpassed 420 ppm, levels that have not been seen for millions of years, marking a clear departure from the natural variability of the Earth's climate (Lindsey, 2021). This unprecedented

156

rise in CO_2 is largely a result of human activities, particularly the widespread combustion of fossil fuels across all sectors of the global economy.

Deforestation and Land-Use Change

Forests play a critical role in mitigating climate change by acting as carbon sinks, absorbing CO_2 from the atmosphere and storing it in biomass and soil. However, deforestation and land-use changes have significantly diminished the Earth's capacity to sequester carbon. Deforestation, particularly in tropical regions, is driven by agricultural expansion, logging, and urbanization. When forests are cleared, either through logging or burning, the carbon stored in trees and soil is released back into the atmosphere, exacerbating the effects of climate change. Additionally, the conversion of forests into agricultural land, particularly for the cultivation of crops like soy and palm oil or for cattle ranching, contributes to long-term carbon emissions through both direct release and the loss of carbon sequestration capacity (Houghton, 2015).

Tropical deforestation is a major contributor to the disruption of the global carbon cycle, as these regions historically absorbed large amounts of CO_2. The loss of these vital carbon sinks accelerates climate change by reducing the Earth's ability to balance the carbon budget. Furthermore, the conversion of forests to agricultural land introduces new sources of emissions, as the soil that was once rich in organic carbon is often disturbed and exposed to oxygen, releasing additional CO_2. These processes have not only amplified the impact of fossil fuel emissions but also disrupted ecosystems and biodiversity, compounding the negative effects on the global climate (Houghton, 2015).

Agricultural Practices

Agriculture is responsible for a significant portion of global greenhouse gas emissions, driven primarily by livestock production and the use of synthetic fertilizers. Livestock, particularly cattle, produce methane (CH_4), a potent greenhouse gas, through a process called enteric fermentation in their digestive systems. Methane has a much higher global warming potential than CO_2, with a warming potential 25 times greater over a 100-year period (Smith et al., 2014). Rice paddies also contribute to methane emissions, as the anaerobic conditions in flooded fields promote the production of methane by microorganisms. The global demand for meat and dairy products, particularly in rapidly developing economies, has led to an increase in livestock populations, further compounding methane emissions (Smith et al., 2014).

Additionally, the use of synthetic fertilizers in agriculture has contributed to emissions of nitrous oxide (N_2O), another potent greenhouse gas. Nitrous oxide is released when nitrogen-based fertilizers are applied to crops, and its global warming potential is 298 times greater than that of CO_2 over a 100-year period (UNEP, 2021). The growing demand for food to support the expanding global population has led to increased use of fertilizers and other chemicals in agriculture, further exacerbating the strain on the planet's climate systems. As agricultural practices intensify to meet the needs of a growing global population, the resulting emissions from livestock, rice production, and fertilizer use continue to significantly impact the Earth's carbon cycle, reinforcing the need for sustainable farming practices to mitigate climate change.

Industrial Processes and Waste Management

Apart from energy production, industrial processes such as cement manufacturing contribute significantly to greenhouse gas emissions. Cement is one of the most essential building materials globally, and its production is energy intensive. The process involves both the combustion of fossil fuels to generate the heat needed for the chemical reactions and the chemical transformation of limestone ($CaCO_3$) into lime (CaO), a process known as calcination. When limestone is heated, it releases carbon dioxide (CO_2) as a byproduct of the breakdown of calcium carbonate ($CaCO_3$) into calcium oxide (CaO) and CO_2. This reaction, which occurs at high temperatures in cement kilns, accounts for approximately 60% of the CO_2 emissions from cement production (Andrew, 2018). The remaining emissions come from the burning of fossil fuels, primarily coal, oil, and natural gas, which provide the energy needed to reach the temperatures necessary for calcination.

In total, cement production is responsible for about 7% of global CO_2 emissions, making it one of the largest industrial sources of greenhouse gases worldwide (Andrew, 2018). The demand for cement has grown with the expansion of urban areas, infrastructure projects, and construction industries. This global demand for cement continues to drive emissions higher, and with limited advancements in carbon capture technologies for the cement industry, it remains a major challenge in the effort to mitigate climate change.

Furthermore, the management of solid waste, particularly in landfills, contributes to greenhouse gas emissions, particularly methane (CH_4), which is a potent greenhouse gas with a global warming potential 25 times greater than that of CO_2 over a 100-year period (United Nations Environment Programme [UNEP], 2021). In landfills, organic waste such as food scraps, yard clippings, and paper decompose anaerobically, meaning without oxygen. In the absence of oxygen, bacteria break down the organic material, producing methane as a byproduct. As landfills continue to expand to accommodate increasing urban populations and the growing volume of waste, the amount of methane released into the atmosphere also rises. In fact, landfills are the third-largest source of methane emissions in the United States, contributing significantly to overall greenhouse gas levels (Environmental Protection Agency [EPA], 2020).

As urban populations grow and consumption patterns change, the volume of waste generated continues to increase, which in turn amplifies the methane emissions from landfills. In many developing countries, waste management infrastructure is insufficient, and waste is often disposed of in open dumps, where it decomposes in uncontrolled conditions, further increasing methane emissions. In industrialized nations, while modern landfills are often equipped with methane capture systems, not all facilities have the technology to prevent the release of methane, and some still leak into the atmosphere.

Moreover, waste management practices, including the treatment of organic waste, have become a critical issue in reducing greenhouse gas emissions. Innovations such as composting, anaerobic digestion, and waste-to-energy technologies are increasingly being explored as alternatives to traditional landfill disposal, as these methods can mitigate methane emissions and reduce the carbon footprint of waste management (EPA, 2020). For example, composting organic waste promotes aerobic decomposition, which releases far less methane than anaerobic processes. Similarly, anaerobic digestion can capture methane and use it for energy, turning a potent greenhouse gas into a useful resource.

However, even with advancements in waste management technologies, the growing volume of waste due to population growth, increased consumption, and urbanization presents a significant challenge in managing emissions from industrial processes and waste. As urban areas continue to expand and waste production increases, addressing emissions from industries like cement manufacturing and managing waste through more sustainable practices will be essential to mitigating the overall impact of human activities on the climate system.

Evidence of Human Impact on Recent Warming

The evidence of human impact on recent warming is both overwhelming and irrefutable. Over the past century, global temperatures have risen at an unprecedented rate, with human activities, particularly the burning of fossil fuels, deforestation, and industrial processes, playing a central role in driving these changes. As the planet's climate systems have shifted, the consequences have become increasingly apparent, from rising sea levels and melting ice caps to more frequent and intense extreme weather events. These signs not only confirm the reality of climate change but also underscore the urgent need for action to address the human-driven factors that are accelerating the planet's warming trajectory.

Rising Global Temperatures

The most apparent evidence of human-driven climate change is the rise in global temperatures. Over the last century, the Earth's average temperature has increased by approximately 1.1°C, with this warming becoming particularly pronounced in recent decades (IPCC, 2021). This temperature increase is not uniform across the globe, with significant regional disparities emerging. Some regions, particularly in the Arctic, have experienced warming rates more than twice the global average, a phenomenon known as *Arctic amplification* (Serreze & Barry, 2011). This accelerated warming in the polar regions is driven by the loss of reflective sea ice, which once helped to cool the Earth by reflecting solar radiation. As ice melts and exposes darker ocean surfaces, more heat is absorbed, further intensifying warming in these sensitive areas (Screen & Simmonds, 2010).

These changes in temperature have had far-reaching impacts on both natural ecosystems and human societies. Shifting temperature patterns are altering the timing and behavior of seasonal events, such as plant blooming and animal migration, disrupting established ecosystems. Additionally, human societies have felt the effects through altered agricultural productivity, with some regions experiencing longer growing seasons while others suffer from droughts and extreme heat, all of which threaten food security and livelihoods. The warming of the Earth also exacerbates existing environmental problems, such as desertification and the spread of diseases, making it increasingly challenging to adapt to these rapid changes (IPCC, 2021).

Melting Ice and Rising Seas

The warming of the Earth's climate has led to widespread melting of glaciers and polar ice caps, providing stark evidence of the ongoing climate crisis. Since the 1970s, the extent of Arctic sea ice has decreased by more than 40%, with some areas experiencing ice loss that exceeds 50% during summer months (NASA, 2020). This dramatic reduction in ice coverage is a direct result of rising temperatures in the region, which have been amplified by feedback mechanisms, such as the loss of albedo as ice melts and exposes darker ocean water that absorbs more heat (Screen & Simmonds, 2010). In addition to sea ice decline, glaciers and ice sheets in both Greenland and Antarctica are losing mass at accelerating rates, contributing to rising sea levels. The melting of these large ice masses has been one of the primary drivers of the observed global sea-level rise, which has increased by about 20 cm since 1900, with projections suggesting that sea levels could rise by an additional 1 meter or more by 2100, depending on future emissions scenarios (IPCC, 2021).

This rising sea level poses a direct and immediate threat to coastal communities and ecosystems around the world. Low-lying areas such as island nations and coastal cities are at risk of flooding and erosion, with millions of people potentially displaced due to rising waters. In addition to the physical displacement of people, rising seas also disrupt ecosystems, particularly those in coastal wetlands and estuaries, which provide essential services such as habitat for wildlife, water filtration, and storm surge protection. The effects of rising sea levels on global infrastructure, particularly in cities with dense populations and valuable infrastructure, are profound, requiring significant investment in mitigation measures such as sea walls, flood defenses, and managed retreat.

Extreme Weather Patterns

In addition to the gradual increase in global temperatures, human-induced climate change has been closely linked to an increase in the frequency, intensity, and severity of extreme weather events. Hurricanes, heatwaves, and droughts have become more frequent and severe, serving as stark reminders of the changing climate. For instance, events like Hurricane Katrina in 2005 and the California wildfires in 2020 have highlighted how extreme weather events, fueled by rising temperatures and altered weather patterns, are becoming more destructive and widespread (National Oceanic and Atmospheric Administration [NOAA], 2020). The increased intensity of hurricanes is linked to warmer ocean temperatures, which provide more energy for these storms, resulting in stronger winds and more precipitation. Similarly, heatwaves have become more frequent, especially in regions that were not traditionally prone to extreme heat, exacerbating heat-related illnesses and placing additional strain on energy systems as demand for cooling increases.

Droughts are also becoming more common and prolonged in many regions, with notable impacts on agriculture, water supply, and human livelihoods. In some areas, such as the American West and parts of sub-Saharan Africa, drought conditions have intensified, leading to crop failures, water scarcity, and reduced agricultural productivity (IPCC, 2021). Conversely, other regions have seen increased rainfall and flooding, as changes in atmospheric circulation patterns lead to more intense precipitation events. These shifting precipitation patterns are contributing to a rise in flood events, which, combined with rapid urbanization, have resulted in greater vulnerability to property damage, infrastructure loss, and displacement.

The global distribution of extreme weather events also reflects the uneven nature of climate change impacts. While some regions face prolonged droughts and desertification, others experience heavier rainfall and devastating floods. These changes in weather patterns have significant implications for water management, food security, and human health, as populations face new and unforeseen challenges related to climate change. As global temperatures continue to rise, the frequency and severity of extreme weather events are expected to intensify, highlighting the urgent need for both mitigation and adaptation strategies to address these growing risks.

Climate Feedback Loops Amplified by Human Activity

Climate feedback loops are natural processes that can either amplify or dampen the effects of climate change. However, human activities have significantly intensified these feedback mechanisms, accelerating the rate of global warming. As we alter the planet's land, water, and atmospheric systems, we trigger a cascade of responses that exacerbate the original warming, creating a vicious cycle that is increasingly difficult to reverse. From the loss of ice cover in the Arctic to the release of greenhouse gases from thawing permafrost, these feedback loops are amplifying the impact of human-driven climate change and heightening the urgency for immediate action to mitigate further damage to the planet.

Arctic Amplification

One of the most pronounced and concerning feedback mechanisms driving accelerated climate change is *Arctic amplification*. In this phenomenon, the warming observed in the polar regions occurs at a significantly faster rate than the global average. Since the 1970s, the Arctic has warmed at approximately twice the rate of the global average, and in some parts of the region, the rate is even higher (Serreze & Barry, 2011). This amplified warming is largely due to the loss of sea ice, which decreases the Earth's albedo (reflectivity), allowing more sunlight to be absorbed by the darker ocean surface rather than being reflected back into space. In a naturally icy environment, much of the incoming solar radiation is reflected by the ice, but as the ice melts, the exposed water absorbs more heat, which accelerates regional warming (Screen & Simmonds, 2010).

The reduction of sea ice also contributes to the destabilization of the climate system by altering atmospheric and oceanic circulation patterns, which further accelerate local warming. As the ice melts, the surface temperature of the Arctic ocean rises, which in turn leads to more sea ice loss. This positive feedback loop is self-reinforcing— more ice melts, and in turn, more heat is absorbed, exacerbating the process. The loss of ice also disrupts ecosystems that depend on the ice, such as polar bear habitats and the food chain in Arctic waters (Screen & Simmonds, 2010). This accelerated warming has global consequences, including altering weather patterns, influencing the jet stream, and contributing to the destabilization of other environmental systems far beyond the Arctic region.

Permafrost Thaw

The thawing of permafrost is another significant and worrying feedback loop that is accelerating global warming. Permafrost is ground that remains frozen year-round, typically found in high-latitude regions such as Siberia, Alaska, and northern Canada. This frozen ground stores large quantities of organic matter, including carbon and methane, which have been trapped for thousands of years. As the climate warms, permafrost begins to thaw, releasing these greenhouse gases into the atmosphere. The released carbon dioxide (CO_2) and methane (CH_4) contribute to further global warming, creating a feedback loop that amplifies climate change (Schuur et al., 2015).

Methane, in particular, is a potent greenhouse gas, with a global warming potential more than 25 times greater than CO_2 over a 100-year period (United Nations Environment Programme [UNEP], 2021). The potential for a rapid release of methane from thawing permafrost is particularly concerning because it could lead to a "tipping point" where the release of gases becomes self-sustaining and uncontrollable. As the thawing process accelerates, it not only releases carbon stored in the soil but also destabilizes the land itself, leading to shifts in ecosystems, infrastructure damage, and increased risks of wildfires. This feedback loop is one of the most concerning because the gases released could significantly alter atmospheric composition and potentially push the planet toward more extreme climate scenarios (Schuur et al., 2015).

Oceanic Changes

The warming of the world's oceans represents another critical feedback loop with profound implications for global climate systems. As atmospheric temperatures rise, ocean temperatures increase as well, leading to several cascading effects. One of the most visible impacts of warmer oceans is coral bleaching. Coral reefs, which rely on specific temperature ranges to survive, are highly sensitive to even slight increases in water temperature. When ocean temperatures exceed this threshold, corals expel the symbiotic algae that live within their tissues, causing them to lose their vibrant color and potentially die. This disrupts marine ecosystems that rely on coral reefs for habitat, food, and shelter, with devastating consequences for biodiversity (Hoegh-Guldberg et al., 2018).

In addition to coral bleaching, warmer oceans also have broader impacts on marine ecosystems and food chains. The warming of the seas alters the distribution of marine species, disrupts migration patterns, and affects the productivity of key species, such as plankton, which form the base of the ocean food chain (Hoegh-Guldberg et al., 2018). Furthermore, increased carbon dioxide (CO_2) in the atmosphere is absorbed by the oceans, leading to ocean acidification. The absorption of CO_2 lowers the pH of seawater, making it more acidic and threatening organisms that rely on calcium carbonate to form shells, such as shellfish and coral. Ocean acidification disrupts marine food webs and threatens the livelihoods of millions of people who rely on fishing and aquaculture for food and income (Hoegh-Guldberg et al., 2018).

Another key aspect of oceanic change is the disruption of ocean circulation patterns, such as the thermohaline circulation, also known as the "global conveyor belt." This circulation system, which drives heat exchange between the equator and the poles, is powered by differences in water density, which is affected by temperature and salinity. As ocean temperatures rise and ice melts, the influx of freshwater from melting glaciers can alter the salinity and density of seawater, weakening the thermohaline circulation. This disruption can have significant global impacts, including changes in weather patterns, particularly in Europe, which is heavily influenced by the Gulf Stream, and shifts in the distribution of nutrients in the oceans, affecting marine life (Broecker, 1991). The disruption of these oceanic processes contributes to a feedback loop that amplifies the effects of climate change, further altering the planet's climate systems.

The Anthropocene and the Tipping Points

The concept of the Anthropocene marks a new era in Earth's history, where human activity has become the dominant force shaping the planet's systems. From the rapid acceleration of climate change to widespread biodiversity loss and the transformation of landscapes, human actions have left an indelible mark on the environment. Central to this epoch is the concept of tipping points, critical thresholds beyond which certain environmental changes become irreversible and self-perpetuating. As humanity continues to push the Earth's systems beyond their natural limits, these tipping points represent the greatest risks to the stability of the global climate, ecosystems, and the future of human civilization itself. Understanding and addressing these tipping points is crucial if we are to mitigate the long-term consequences of our actions and avoid reaching points of no return.

The Anthropocene Epoch

The concept of the Anthropocene represents a profound shift in Earth's geological history, signaling an era where human activity has emerged as the dominant force shaping the planet's systems. Proposed by atmospheric chemist Paul Crutzen in the early 2000s, the Anthropocene describes a new geological epoch characterized by humanity's profound and far-reaching impact on the Earth (Crutzen, 2002). Over the past few centuries, human activity has drastically altered the Earth's climate, ecosystems, and natural processes, particularly through the burning of fossil fuels, deforestation, urbanization, and industrial agriculture. The most visible and dramatic effect of this new epoch is the rise in atmospheric greenhouse gases, particularly carbon dioxide (CO_2), methane (CH_4), and nitrous oxide (N_2O), which have led to global warming and climate change at an unprecedented pace (IPCC, 2021).

However, the Anthropocene is not defined solely by climate change. It encompasses a broader range of environmental transformations, including widespread biodiversity loss, the alteration of biogeochemical cycles, and the depletion of natural resources. Human activities, such as agriculture, urbanization, and industrial production, have not only changed the composition of the atmosphere but also disrupted the natural cycling of key elements like carbon, nitrogen, and phosphorus, leading to significant ecological imbalances (Steffen et al., 2015). The massive loss of species, driven by habitat destruction, over-exploitation, and pollution, is another hallmark of the Anthropocene. Current extinction rates are estimated to be hundreds to thousands of times higher than the natural background rate, signaling the sixth mass extinction event in Earth's history (Ceballos et al., 2015).

Furthermore, humanity's unrelenting extraction of natural resources, whether through mining, deforestation, or water usage, has altered the planet's physical landscape, depleting ecosystems and pushing many species to the brink of extinction. These transformations are not just localized but have global consequences, as changes in one part of the planet can have cascading effects across regions, further altering the delicate balance of Earth's systems. As a result, the Anthropocene represents not only a period of unprecedented human influence but also a time of significant ecological, atmospheric, and geological disruption, which has profound implications for the future of life on Earth.

Tipping Points and Irreversible Changes

One of the most pressing concerns of the Anthropocene is the potential for reaching tipping points, critical thresholds in the Earth's climate system where small changes can trigger large, irreversible shifts.

These tipping points represent points beyond which natural processes or systems become self-perpetuating, often leading to significant and long-lasting changes to the environment. The risk of surpassing these thresholds increases as human activities continue to stress the planet's systems, driving changes in the climate and ecosystems at an accelerated rate.

One of the most widely discussed tipping points is the collapse of the Amazon rainforest, often referred to as the "lungs of the Earth" due to its role in absorbing carbon dioxide and producing oxygen. The Amazon is not only a critical carbon sink but also a rich biodiversity hotspot, home to millions of species. However, deforestation, climate change, and land degradation have already caused significant stress on this ecosystem. If deforestation continues at current rates or if the climate warms enough to shift rainfall patterns, the Amazon could reach a tipping point where it transitions from a lush, carbon-absorbing rainforest to a degraded savannah-like ecosystem. This shift would not only release vast amounts of CO_2 into the atmosphere but would also dramatically alter global weather patterns, impacting agricultural production and food security (Lenton et al., 2019).

Another key tipping point involves the disintegration of the polar ice sheets, particularly in Greenland and Antarctica. The melting of these ice sheets has already contributed to rising sea levels, and as temperatures continue to rise, the rate of ice loss is expected to accelerate. If these ice sheets reach a critical point of instability, they could collapse irreversibly, leading to significant long-term sea-level rise that would inundate coastal cities and displace millions of people. The melting of polar ice also amplifies Arctic amplification, where the loss of ice reduces the Earth's albedo, absorbing more heat and further accelerating global warming (Serreze & Barry, 2011).

These tipping points could trigger long-term disruptions in climate patterns, posing a severe threat to the stability of ecosystems and human societies. The destabilization of key ecosystems such as the Amazon or the polar ice sheets could lead to cascading effects on water resources, food security, and biodiversity, making adaptation and mitigation more difficult. Once these tipping points are crossed, it may be impossible to reverse the damage, leaving future generations to cope with the consequences of the changes set in motion by human activity. This highlights the urgent need for comprehensive global action to prevent further destabilization of Earth's systems and to safeguard the future of life on the planet.

Future Projections and the Role of Mitigation

Future projections of climate change reveal a range of potential outcomes, heavily dependent on the actions taken in the coming decades. These projections, grounded in various emission scenarios, highlight the significant impact human activities will continue to have on the global climate. While the trajectory of warming can still be influenced, the window for effective action is rapidly closing. Mitigation—through reducing greenhouse gas emissions, transitioning to renewable energy sources, and implementing sustainable practices—will be crucial in shaping the future climate. This section explores the projected outcomes under different emissions pathways and emphasizes the urgent need for coordinated global efforts to mitigate the most devastating effects of climate change.

Projected Warming Scenarios

Future climate projections based on various emission scenarios paint a concerning picture of the planet's trajectory. These projections, developed by the Intergovernmental Panel on Climate Change (IPCC), outline different pathways depending on future greenhouse gas emissions and mitigation efforts. Under the Representative Concentration Pathway (RCP) 8.5 scenario, which assumes continued high emissions with no significant reductions in fossil fuel consumption, global temperatures are projected to rise by as much as 4.8°C by the end of the century (IPCC, 2021). Such a dramatic increase would result in devastating consequences for ecosystems, human health, agriculture, and infrastructure. Sea-level rise, more intense heatwaves, and extreme weather events would become the norm, placing immense strain on societies, especially in vulnerable regions.

In contrast, the RCP 2.6 scenario represents an idealized pathway in which urgent and drastic mitigation measures are implemented globally. This scenario envisions a significant transition to renewable energy, widespread adoption of energy-efficient technologies, and substantial reductions in greenhouse gas emissions. If the global community can achieve these ambitious goals, the temperature rise could be limited to below 2°C by the end of the century. While even 2°C of warming would still present challenges, it would allow for a more manageable adjustment period and could help avoid the most catastrophic outcomes, such as the irreversible loss of ecosystems or the collapse of critical natural systems (IPCC, 2021). The stark contrast between these two scenarios underscores the critical importance of implementing effective policies and actions that drastically reduce emissions and promote the transition to sustainable energy sources.

Mitigation Efforts and Challenges

Mitigation efforts are essential for limiting future warming and curbing the most harmful effects of climate change. The transition to renewable energy sources, such as solar, wind, and hydroelectric power, is central to reducing global emissions, as the energy sector is responsible for a substantial portion of global greenhouse gas emissions. Carbon capture technologies, which aim to capture CO_2 emissions at their source and store them underground or use them for other purposes, represent another potential strategy for reducing atmospheric carbon levels. Additionally, improving energy efficiency in buildings, transportation, and manufacturing can significantly reduce emissions by lowering the overall demand for energy.

However, achieving a low-carbon economy is fraught with challenges. Political resistance remains one of the largest barriers to effective climate action. In many countries, there is opposition to policies that reduce fossil fuel use due to vested interests in industries such as coal, oil, and gas, as well as fears of economic disruption (Stern, 2007). Technological barriers also pose a challenge, as many green technologies are not yet as efficient or cost-effective as traditional fossil fuel-based options. The implementation of carbon capture and storage technologies, for example, remains in its early stages and faces both technical and financial hurdles that must be overcome to make it viable at a global scale. Economic considerations are another key challenge, as the transition to renewable energy and sustainable practices requires substantial investment in infrastructure, research, and development. While the long-term benefits of these investments are clear, the upfront costs are significant and often require government intervention and international cooperation to secure funding and ensure equitable distribution.

International agreements like the Paris Agreement have been instrumental in creating a global framework for climate action, with nearly every nation agreeing to reduce emissions and limit global warming. However, progress toward these goals has been slow and uneven, with many countries failing to meet their emissions targets or delay implementing necessary measures. The Paris Agreement calls for the global community to limit warming to well below 2°C, with an ambition to limit it to 1.5°C. However, current emissions trends indicate that the world is on track for a much higher level of warming without urgent and sustained efforts (UNFCCC, 2015). To meet these targets, it will require unprecedented international cooperation, large-scale investments in green technologies, and significant shifts in global consumption patterns.

The Role of Individuals and Communities

While large-scale systemic change is crucial for mitigating climate change, individual actions and community-level efforts also play a vital role. Grassroots movements and local initiatives can significantly reduce emissions, promote sustainability, and raise awareness about climate change. For instance, urban farming initiatives help reduce the carbon footprint of food production, while sustainable transportation solutions, such as cycling infrastructure or electric vehicle adoption, can cut emissions from the transportation sector. Additionally, individuals can contribute to reducing emissions by altering their consumption patterns, reducing waste, and supporting policies that encourage environmental sustainability.

Local solutions, such as community-based renewable energy projects, can also help reduce reliance on fossil fuels. By adopting sustainable practices at the local level, communities can help drive broader societal changes, particularly when these efforts are supported by policy frameworks that incentivize green practices. In many cases, the collective actions of individuals and communities can create powerful networks that challenge the status quo and advocate for systemic change. The role of individuals in the fight against climate change cannot be underestimated, as even small lifestyle changes, such as reducing energy consumption, adopting plant-based diets, and supporting eco-friendly businesses, can have a significant collective impact when scaled up across entire populations (Hawken, 2017).

Humanity's Path Forward

Humanity stands at a pivotal crossroads, where the future of the planet hangs in the balance. The immense toll we've exacted on the Earth, through industrialization, deforestation, and unsustainable consumption, has set in motion profound disruptions to our climate and ecosystems. Yet, amid these daunting challenges, there is hope. The power to change course lies in our hands. Through unwavering global cooperation, groundbreaking technological innovation, and bold individual action, we can halt the most destructive trends and carve out a path toward a sustainable future. The urgency of the moment cannot be overstated, our planet's survival depends on our collective resolve to act, to innovate, and to rise above inertia. By embracing sustainability, resilience, and transformative change, we can not only mitigate the impacts of climate change but create a world where future generations can thrive, in harmony with a healthier, more resilient Earth. The time to act is now, there is no more room for delay.

Chapter 16

Living with Change

Climate change is no longer a distant possibility, its impacts are being felt today, shaping our weather patterns, ecosystems, and societies. While efforts to halt climate change are critical, the complexities of Earth's climate system, combined with the momentum of ongoing human-induced changes, make the task of fully reversing it impossible. The greenhouse gases already emitted into the atmosphere have set in motion long-term changes that cannot be undone in the near future (IPCC, 2021). As a result, the focus must shift from simply trying to prevent climate change to a more comprehensive strategy that includes reducing our impact, adapting to the inevitable changes already underway, and mitigating their effects for a sustainable future. Accepting this reality does not mean giving up on climate action, but rather recognizing that the pathway forward will require resilience, innovation, and a commitment to making the most of the time we have to shape a livable future for all.

The Irreversibility of Certain Climate Impacts

Certain impacts of climate change are now irreversible, marking a critical shift in our understanding of the planet's future. The extensive damage already done to key ecosystems and climate systems means that some changes are locked in, regardless of future actions to reduce emissions or mitigate further damage. This includes the loss of ice in polar regions, the degradation of coral reefs, and shifts in biodiversity that cannot be undone in the short term. These irreversible effects underscore the urgency of addressing climate change not only by reducing emissions but also by adapting to the changes that are already set in motion, acknowledging that some of the most profound shifts in the Earth's climate are beyond our ability to reverse.

Lagging Effects in the Climate System

One of the most significant challenges in addressing climate change lies in the lag in the climate system's response to greenhouse gas emissions. The Earth's climate operates on timescales that can span decades to centuries, meaning the full impact of current emissions will not be felt for years or even centuries. Greenhouse gases such as carbon dioxide (CO_2), methane (CH_4), and nitrous oxide (N_2O) persist in the atmosphere for varying lengths of time, with CO_2 having a particularly long atmospheric lifetime, ranging from 100 to 1,000 years (Lashof & Ahuja, 1990). Even if global emissions were halted immediately, the climate would continue to warm for many years due to the greenhouse gases already in the atmosphere. This "commitment to warming" means that the planet is already experiencing a level of warming that is a consequence of past emissions, and the climate system will continue to evolve based on the greenhouse gases that remain in the atmosphere.

The persistence of these gases in the atmosphere acts like a blanket around the Earth, trapping heat and raising global temperatures. This phenomenon, known as the greenhouse effect, is a primary driver of climate change. The warming we are experiencing now is not just the result of emissions from the past few years, but also from decades of industrialization and deforestation, which have released significant amounts of CO_2 and other gases into the atmosphere. As a result, even if immediate global action were taken to reduce emissions to zero, the

Earth's climate would continue to warm due to the cumulative effects of past emissions (IPCC, 2021). This lag is compounded by the slow response of climate systems, including the oceans, which absorb much of the heat and carbon dioxide. As the Earth system continues to adjust, the long-term impacts will be felt by future generations, underscoring the deep and enduring nature of climate change.

The implications of this lag are vast. We are already witnessing more frequent and severe weather events, such as heatwaves, droughts, and heavy rainfall, all of which are intensified by the warming that has already occurred. Additionally, ecosystems and species that have adapted to stable conditions over thousands of years are now being forced to adjust to rapidly changing environments. The lag in the climate system highlights the need for long-term strategies not only to reduce future emissions but also to mitigate the inevitable impacts of the warming that has already been set in motion.

Tipping Points Already Passed

Climate scientists have long warned that certain tipping points in the Earth's systems have already been passed, making some changes irreversible. These tipping points refer to thresholds beyond which the climate system undergoes significant, often irreversible, changes. One of the most prominent and concerning examples is the significant reduction in Arctic sea ice. Arctic amplification, a phenomenon where the Arctic warms faster than the global average, has led to dramatic reductions in sea ice cover, a process that has already passed critical thresholds (Serreze & Barry, 2011). The loss of sea ice not only contributes directly to rising sea levels but also exacerbates global warming through a feedback loop. As the ice melts, the surface area of darker ocean water is exposed, which absorbs more solar radiation than the ice, further warming the region and accelerating the melting of ice. This positive feedback loop is difficult to stop and is unlikely to reverse in the near term, even with drastic reductions in emissions. The continued loss of Arctic ice contributes to global sea level rise and alters global weather patterns, further intensifying the effects of climate change (IPCC, 2021).

Another key example of a passed tipping point is the widespread coral bleaching that has occurred due to rising ocean temperatures. Coral reefs, which are some of the most diverse and productive ecosystems on the planet, are under extreme stress due to the warming of ocean waters. When sea temperatures rise beyond the tolerance limits of corals, they expel the symbiotic algae living within their tissues, causing them to bleach and often leading to their death (Hoegh-Guldberg et al., 2018). Coral reefs are already in rapid decline globally, with many reefs experiencing mass bleaching events in recent years. These ecosystems are not only biodiversity hotspots but also provide essential services such as coastal protection, food sources, and tourism revenue. The degradation of coral reefs is a clear signal of a tipping point that has been crossed. The changes in coral reefs are unlikely to be fully reversed, even with drastic reductions in greenhouse gas emissions, highlighting the urgency of preserving what remains of these ecosystems and implementing strategies to protect other vulnerable systems.

Other tipping points, such as the destabilization of the Amazon rainforest and the thawing of permafrost in the Arctic, represent further examples of irreversible changes that have already begun to unfold. The Amazon, once considered a vital carbon sink, is now under significant threat from deforestation and climate change. If the forest continues to degrade, it could transition from a carbon sink to a carbon source, releasing vast amounts of CO_2 into the atmosphere and exacerbating global warming. Similarly, the thawing of permafrost is releasing methane, a potent greenhouse gas, into the atmosphere, further accelerating climate change (Schuur et al., 2015). These tipping points, once crossed, are difficult to reverse and may have long-lasting consequences for the stability of Earth's climate and ecosystems.

The passing of these tipping points underscores the critical need for not only reducing greenhouse gas emissions but also prioritizing efforts to adapt to the changes that have already been set in motion. While some changes are irreversible, efforts to mitigate their impacts and protect vulnerable ecosystems and communities remain essential to limiting further damage and creating a sustainable future.

Reducing Greenhouse Gas Emissions to Slow the Pace

Reducing greenhouse gas emissions is the most effective strategy we have to slow the pace of climate change and avoid the worst possible outcomes. While some climate impacts are already locked in, significantly cutting emissions today can help mitigate further warming and reduce the severity of future disruptions. Transitioning to renewable energy sources, improving energy efficiency, and adopting low-carbon technologies are essential steps in reducing emissions from key sectors such as energy, transportation, and industry. These efforts not only help slow the rate of global warming but also provide opportunities for innovation, economic growth, and job creation in the emerging green economy. In order to keep global temperature rise within manageable levels, substantial and immediate action is required to drastically reduce emissions and transition to a sustainable, low-carbon future.

Transitioning to Renewable Energy

One of the most effective and scalable strategies to combat climate change is the transition to renewable energy sources. The widespread adoption of wind, solar, and hydroelectric power is essential to reducing our reliance on fossil fuels such as coal, oil, and natural gas, which are the primary contributors to greenhouse gas emissions. Over the last few decades, technological advances have made renewable energy sources not only more accessible but also far more cost-effective than ever before. Solar energy, for instance, has seen a dramatic reduction in installation and production costs, making it increasingly competitive with fossil fuels. According to the International Renewable Energy Agency (IRENA), the cost of solar photovoltaic (PV) electricity has fallen by approximately 82% between

2010 and 2020, allowing solar energy to become one of the cheapest sources of power generation worldwide (IRENA, 2020).

Similarly, wind power, particularly offshore wind farms, has experienced significant growth. Offshore wind farms, which are capable of generating much larger amounts of electricity compared to onshore turbines due to stronger and more consistent wind patterns, have become a major focus in regions like Europe and the United States. In Europe, countries such as Denmark, the United Kingdom, and Germany are leading the charge, with ambitious offshore wind projects providing substantial contributions to their renewable energy targets (Lund, 2020). In the U.S., states like California, New York, and Massachusetts are also increasing their investment in offshore wind as a critical component of their energy transition strategies.

The global shift toward renewable energy sources is pivotal in reducing carbon emissions, which are the primary drivers of global warming. By shifting from fossil fuel-based power generation to clean, renewable alternatives, countries can significantly decrease their carbon footprints and slow the rate of climate change. This transition will not only help to mitigate global warming but also reduce other environmental hazards associated with fossil fuel extraction and burning, such as air and water pollution. Furthermore, renewable energy development offers opportunities for job creation, economic growth, and enhanced energy security, as countries become less dependent on volatile fossil fuel markets (IEA, 2020).

Energy Efficiency and Conservation

In addition to the widespread adoption of renewable energy, improving energy efficiency across all sectors of the economy is a critical strategy for reducing emissions. Energy efficiency refers to using less energy to perform the same task, and it plays a vital role in curbing global carbon emissions. One of the most effective ways to improve energy efficiency is by retrofitting buildings. Buildings, which account for approximately 40% of global energy use, can be made more energy-efficient by upgrading insulation, installing energy-efficient windows, and improving heating, ventilation, and air conditioning (HVAC) systems. Additionally, adopting energy-efficient appliances, such as LED lighting and Energy Star-rated appliances, can significantly reduce energy consumption in residential, commercial, and industrial settings.

The transportation sector is another key area where efficiency improvements can yield significant emissions reductions. Road transport is responsible for nearly 14% of global greenhouse gas emissions (IEA, 2020), and improving the efficiency of vehicles can substantially reduce these emissions. The transition to electric vehicles (EVs) is one of the most promising solutions, as EVs have zero tailpipe emissions and can be powered by renewable energy sources. With advancements in battery technology, the range and affordability of EVs have improved, making them a more viable option for individuals and businesses alike. Furthermore, the development of public transit infrastructure, including electric buses and trains, can further reduce emissions from the transportation sector by providing more sustainable alternatives to private vehicle use.

Energy conservation, which focuses on reducing overall energy consumption, also plays a critical role in mitigating climate change. This can be achieved through simple practices like reducing unnecessary lighting, optimizing industrial processes to use less energy, and implementing energy-saving technologies in manufacturing. Reducing water consumption also plays a part, as water treatment and heating processes require significant amounts of energy. Encouraging individuals and businesses to adopt energy conservation practices can reduce the overall demand for energy and lower emissions, complementing efforts to transition to renewable energy sources.

Combined, improvements in energy efficiency and the adoption of energy conservation practices can significantly slow the pace of climate change. These strategies complement the broader transition to renewable energy, making it more feasible to meet global emission reduction targets while ensuring economic stability and energy security. By focusing on energy efficiency across all sectors, we can create a more sustainable and resilient energy system, one that meets the needs of the present without compromising the ability of future generations to thrive.

Carbon Capture and Sequestration

Carbon capture and sequestration (CCS) is a critical technology in the fight against climate change, designed to capture carbon dioxide (CO_2) emissions from industrial processes and power generation before they can enter the atmosphere. By storing CO_2 underground or using it in other processes, CCS has the potential to significantly reduce greenhouse gas concentrations in the atmosphere, mitigating some of the impacts of climate change. While the technology holds promise, its widespread implementation faces numerous challenges, including high costs, infrastructure requirements, and the need for rigorous monitoring to ensure the safe and permanent storage of carbon. As part of a comprehensive climate strategy, CCS could play a crucial role in achieving global emissions reduction targets, particularly for sectors that are difficult to decarbonize, such as heavy industry and certain power generation processes.

Nature-Based Solutions

Nature-based solutions (NbS) are an increasingly recognized and effective approach to mitigating climate change by harnessing the power of natural processes to absorb and store carbon dioxide (CO_2) from the atmosphere. Forests are one of the most significant and well-known carbon sinks, absorbing vast amounts of CO_2 through photosynthesis and storing it in their biomass (trees, plants) and soils. Forests are estimated to store roughly 30% of the global land-based carbon pool (Lal, 2005). However, deforestation and forest degradation have greatly reduced the capacity of forests to act as carbon sinks, while simultaneously releasing large amounts of stored carbon back into the atmosphere. Deforestation, driven by agricultural expansion, logging, and urban development, accounts for a significant proportion of global CO_2 emissions.

Reforestation projects, which aim to restore degraded or deforested lands by planting trees, are essential for reversing some of the damage done to global carbon storage capacity. These projects can also enhance biodiversity, improve water quality, and prevent soil erosion. In addition to reforestation, afforestation, the planting of trees on land that has not been previously forested, can contribute significantly to carbon sequestration. By increasing the amount of forest cover, reforestation and afforestation projects help not only absorb more CO_2 but also restore ecosystems that provide a host of other ecosystem services, such as habitat for wildlife, water filtration, and local climate regulation.

Wetland restoration is another vital nature-based solution that holds considerable promise for both carbon sequestration and broader environmental benefits. Wetlands, including marshes, swamps, and mangroves, store large amounts of carbon in their soils due to the slow decomposition of organic material in waterlogged conditions. Wetlands are among the most efficient ecosystems for carbon storage, with some estimates suggesting that they account for up to one-third of the Earth's carbon stocks despite covering only 6% of the land surface (Miller, 2019). Restoring wetlands that have been drained or degraded for agriculture, development, or resource extraction can help store additional carbon and protect against the release of stored carbon. Moreover, wetlands provide critical ecosystem services such as flood mitigation, water purification, and coastal protection from storm surges, making their restoration a win-win strategy for both climate change mitigation and resilience (Miller, 2019).

Incorporating NbS into climate change strategies not only helps reduce atmospheric CO_2 but also strengthens ecosystem resilience and improves human well-being. These solutions offer cost-effective and scalable ways to combat climate change while enhancing biodiversity and ecosystem services that are vital for sustaining life on Earth.

Technological Innovations

While nature-based solutions offer significant benefits, technological innovations are also being developed to complement these efforts by directly capturing and removing CO_2 from the atmosphere. One such innovation is direct air capture (DAC), a process in which CO_2 is captured from ambient air and either stored underground or converted into useful products such as synthetic fuels or building materials. DAC has the potential to play a critical role in mitigating climate change by removing CO_2 that has already been emitted, a necessary component of any strategy to reduce global temperatures.

The technology behind DAC is still in its early stages but holds significant promise for large-scale deployment. It works by using chemical processes to capture CO_2 from the air, after which the gas is compressed and transported to storage sites, typically underground, where it can be safely stored in geological formations, such as depleted oil and gas fields or deep saline aquifers (Keith et al., 2018). DAC technologies can be deployed anywhere, which makes them particularly attractive as a tool for offsetting emissions from hard-to-decarbonize sectors, such as aviation, cement production, and steel manufacturing, where direct emissions reductions are more challenging.

However, the large-scale deployment of DAC faces several significant challenges. One of the primary obstacles is cost. Currently, DAC technologies are expensive, with estimates for the cost of capturing one ton of CO_2 ranging from \$100 to \$600, depending on the technology and location (Keith et al., 2018). This makes DAC a relatively expensive option compared to other carbon reduction strategies, such as renewable energy adoption or energy efficiency improvements. In addition to high initial costs, DAC also requires a significant amount of energy to operate the systems, raising concerns about whether the energy used in the capture process could offset the benefits of CO_2 removal. The energy source for DAC must therefore be renewable to ensure that it contributes positively to the climate and does not inadvertently increase emissions.

The infrastructure required for the transportation and storage of captured CO_2 is another challenge. Safe and permanent storage of CO_2 requires carefully selected geological formations and rigorous monitoring to prevent leaks. Furthermore, scaling up DAC technology to the level needed to make a meaningful impact on global CO_2 levels will require massive investment in infrastructure, research, and regulatory frameworks to ensure that the technology is deployed safely and effectively.

Despite these challenges, DAC offers significant potential as part of a broader portfolio of climate change mitigation strategies. As the technology matures and costs decrease, it could become an essential tool for reducing atmospheric CO_2 levels. Technological solutions such as DAC are crucial for achieving the ambitious global climate goals, including those set by the Paris Agreement, which seeks to limit global warming to well below 2°C.

Both nature-based solutions and technological innovations offer complementary approaches to addressing climate change. While NbS can be implemented immediately and have multiple co-benefits, technologies like DAC represent a critical piece of the puzzle for long-term carbon removal. Both strategies will need to be pursued simultaneously, alongside efforts to reduce emissions, to effectively combat climate change and avoid the most catastrophic outcomes.

Adapting to the Inevitable

Adapting to the inevitable impacts of climate change is no longer a choice but a necessity. While efforts to mitigate climate change are crucial, certain changes to the environment—such as rising sea levels, more frequent extreme weather events, and shifting agricultural patterns—are already underway and cannot be fully avoided. To ensure the resilience of both natural ecosystems and human societies, adaptation strategies must be implemented across all sectors. This involves rethinking infrastructure, transforming agricultural practices, and designing cities that can withstand the increasing pressures of a changing climate. By planning for these changes, we can reduce vulnerability, protect communities, and create a more sustainable future despite the challenges we face. Adaptation is not about passively accepting the consequences of climate change but actively preparing for a future that incorporates its realities.

Resilient Infrastructure

As the effects of climate change become increasingly inevitable, adapting our infrastructure to withstand its impacts is a critical step toward ensuring the continued functionality and livability of cities and communities. Rising sea levels, more frequent and intense storms, and extreme heat events are already putting significant pressure on existing infrastructure, particularly in coastal areas that are at high risk from storm surges and flooding. One of the most effective strategies for safeguarding these areas is the design and construction of flood-resistant infrastructure. Coastal cities can invest in sea walls, storm surge barriers, and elevated buildings to protect against rising tides and extreme weather events. These measures act as physical defenses, mitigating the potential for catastrophic damage to homes, businesses, and critical infrastructure like transportation networks and power plants.

Retrofitting existing infrastructure to withstand more frequent heatwaves and storms is also essential. Older buildings and infrastructure, which were not designed with the current climate in mind, may need to be upgraded to handle the increased stress from higher temperatures, stronger storms, and flood risks. For example, buildings can be retrofitted with better insulation, more resilient roofing materials, and storm-resistant windows. Infrastructure such as bridges, roads, and utilities must be reinforced to endure the impacts of extreme weather and shifting climate patterns. This proactive approach to retrofitting is not just about minimizing damage; it's also about future-proofing communities and ensuring that infrastructure continues to function effectively as climate change progresses.

Urban planning is another critical aspect of building climate-resilient cities. Incorporating green spaces, such as parks, green roofs, and urban forests, into city planning can help mitigate the urban heat island effect, where cities experience higher temperatures than surrounding rural areas due to concrete, asphalt, and limited vegetation. Green spaces not only cool cities but also improve air quality, enhance biodiversity, and provide recreational spaces for residents. By integrating green infrastructure into urban planning, cities can promote resilience while improving quality of life for their inhabitants. Furthermore, urban forests and parks serve as flood buffers, absorbing excess water during heavy rainfall and helping to reduce the risk of urban flooding. These adaptive strategies are crucial for minimizing the impacts of climate change and ensuring that communities remain livable in a warming world.

Agricultural Adaptation

Agriculture is one of the sectors most vulnerable to the inevitable impacts of climate change. The changes in precipitation patterns, rising temperatures, increased frequency of droughts, and shifts in growing seasons are all affecting crop yields, food security, and rural livelihoods. Climate change threatens not only the quantity of food produced but also the quality, as extreme weather events, such as floods and droughts, can lead to crop failure and reduced productivity. As the climate continues to shift, the ability to grow certain crops in traditional regions may no longer be feasible. Adaptation in agriculture, therefore, is critical to maintaining global food security.

One of the primary strategies for agricultural adaptation is shifting crop patterns to match changing climatic conditions. This may involve planting crops that are better suited to new temperature and precipitation regimes or introducing more resilient varieties of existing crops. For instance, in areas facing water scarcity, crops that require less water or are more drought-tolerant, such as sorghum or millet, may become more viable alternatives to traditional crops like rice or wheat. Shifting planting times or introducing new cropping systems, such as agroforestry or intercropping, can also help adapt to changes in precipitation and growing seasons.

Advances in biotechnology offer promising solutions for adapting agriculture to climate change. Genetically modified (GM) crops that are resistant to heat, drought, and pests are being developed to withstand increasingly challenging growing conditions. For example, genetically engineered crops like drought-tolerant corn and heat-resistant rice are already being tested and have shown promising results in regions affected by extreme weather. These GM crops could help maintain crop yields despite adverse conditions, ensuring a stable food supply in regions vulnerable to climate change. However, the adoption of genetically modified crops must be accompanied by careful consideration of ecological, social, and ethical factors, including potential impacts on biodiversity and the availability of seeds to smallholder farmers.

Water management is another key component of agricultural adaptation. As climate change exacerbates water scarcity in many regions, improving the efficiency of irrigation systems and investing in water conservation measures will be critical to ensuring a stable food supply. Precision agriculture, which involves using advanced technology like sensors, drones, and satellite imagery to optimize water use and reduce waste, offers a way to improve water efficiency in farming. Additionally, rainwater harvesting and the use of recycled wastewater for irrigation can help reduce dependence on overexploited freshwater sources. In regions prone to flooding, incorporating flood-resistant agricultural practices, such as raised beds or flood-tolerant crop varieties, can protect against crop losses due to excessive rainfall.

To further support agricultural adaptation, governments and international organizations must invest in research, education, and extension services to help farmers implement these strategies.

Providing farmers with access to climate information, early warning systems, and training in sustainable farming practices will be essential for building resilience and ensuring long-term food security. As the climate continues to change, it is clear that adaptation strategies in agriculture will play a crucial role in securing the food systems of the future and supporting the livelihoods of millions of people around the world.

Mitigating Social and Economic Impacts

Mitigating the social and economic impacts of climate change is critical to ensuring that the most vulnerable communities are protected and that the global economy can transition to a more sustainable future. Climate change is not just an environmental issue; it has profound social and economic consequences that exacerbate existing inequalities and disrupt livelihoods, particularly in developing regions. Addressing these impacts requires comprehensive strategies that focus on protecting vulnerable populations, diversifying economies reliant on fossil fuels, and providing financial tools, such as climate insurance, to help communities recover from extreme weather events. By adopting inclusive policies that prioritize equity and resilience, we can create a more just and sustainable future, where the burden of climate change is shared fairly, and the most at-risk populations receive the support they need to thrive in an increasingly uncertain world.

Protecting Vulnerable Communities

The impacts of climate change are not felt equally across the globe, with vulnerable communities, particularly those in developing countries, bearing the brunt of its effects. These communities often have the least capacity to adapt to rapidly changing environmental conditions due to limited resources, inadequate infrastructure, and lower levels of economic development. As a result, they are disproportionately affected by climate-related risks such as rising sea levels, extreme heatwaves, droughts, floods, and extreme weather events like hurricanes and cyclones. These climate impacts exacerbate existing inequalities, threatening the livelihoods of millions of people who rely on agriculture, fishing, and other climate-sensitive industries.

For instance, small island nations in the Pacific and Caribbean face existential threats from rising sea levels, which encroach on their land, infrastructure, and freshwater supplies. Similarly, many regions in sub-Saharan Africa are experiencing increased frequency and intensity of droughts, making food and water access increasingly unreliable for local populations. Urban areas in developing nations, often densely populated and under-resourced, are also at high risk from flooding and the urban heat island effect. In these contexts, the most marginalized groups, such as women, children, the elderly, and indigenous populations, are often the hardest hit due to their limited access to resources, social protection, and decision-making power (UNFCCC, 2015).

Addressing these disparities requires concerted efforts from the international community, particularly through climate financing and targeted adaptation programs. Climate financing is crucial for helping developing countries fund adaptation initiatives, such as building flood defenses, improving water management systems, and enhancing agricultural resilience to changing weather patterns. The Green Climate Fund (GCF), established under the UNFCCC, plays a central role in supporting these efforts by providing financial assistance to countries that are most vulnerable to climate impacts. Programs aimed at empowering communities to take action on climate change are also vital, including capacity-building initiatives that enable local communities to assess their vulnerability and implement adaptation strategies that are culturally and contextually appropriate.

Moreover, international cooperation is necessary not just for providing financial support but also for sharing knowledge, technologies, and best practices to build resilience in vulnerable regions. By working together, developed and developing nations can create mechanisms that ensure the most vulnerable populations are not left behind, empowering them to adapt and thrive in a changing climate.

Economic Diversification

As the global economy transitions to a low-carbon future, regions that are heavily dependent on fossil fuels face particular challenges. These communities, which often rely on industries like coal mining, oil extraction, and gas production, are at risk of economic disruption as demand for these energy sources declines due to the global shift towards renewable energy. In regions where fossil fuel industries are a major source of employment and economic activity, the loss of these industries could lead to widespread job losses, economic decline, and social instability. Therefore, economic diversification is essential for ensuring a sustainable future for these communities.

Governments can support this transition by investing in green industries, such as renewable energy, green manufacturing, and sustainable agriculture. Renewable energy industries, such as wind, solar, and hydroelectric power, offer immense opportunities for job creation and economic growth. For example, the renewable energy sector has already become a significant source of employment in countries like Germany and China, with solar and wind energy jobs growing rapidly in response to increased demand. Green manufacturing, which focuses on producing energy-efficient products and technologies, can similarly provide new avenues for growth in regions that traditionally relied on heavy industries. Sustainable agriculture, which includes practices like organic farming, agroforestry, and precision agriculture, can help reduce emissions from agriculture while also improving food security and rural development.

Training programs and educational initiatives are critical to helping workers transition from fossil fuel-dependent industries to the green economy. Programs that provide skills training in renewable energy technologies, energy efficiency, and sustainable farming practices are essential for ensuring that workers in fossil fuel industries can find new employment opportunities. Governments, private sector companies, and educational institutions must collaborate to offer retraining programs that equip workers with the skills needed to thrive in these emerging sectors. Economic diversification efforts should also focus on creating entrepreneurial opportunities, particularly in the green technology and sustainable services sectors, to encourage innovation and local business development.

By promoting economic diversification, governments can help reduce the negative impacts of climate change on employment, create new opportunities for sustainable economic growth, and ensure that communities are more resilient to the transitions required in the low-carbon economy.

Insurance and Risk Management

As climate change continues to increase the frequency and intensity of extreme weather events such as storms, floods, and wildfires, the financial burden on communities and businesses will grow substantially. This is where insurance and risk management play a crucial role in helping individuals and organizations manage and recover from the financial losses associated with these disasters. Insurance provides a safety net for those affected by natural disasters, helping them recover, rebuild, and restore their livelihoods. For businesses, insurance can help protect against the costs of disrupted operations, allowing them to resume activities more quickly.

Expanding climate insurance programs is critical for providing financial protection, especially for communities and regions that are particularly vulnerable to climate impacts. Insurance products designed to cover risks related to floods, droughts, and extreme weather can reduce the financial vulnerability of individuals and businesses. In regions prone to flooding or extreme heat, for example, governments and private insurers can collaborate to offer affordable insurance policies that provide coverage for property damage, crop losses, and infrastructure repair. In some cases, microinsurance, small-scale insurance products targeted at low-income individuals, can be a valuable tool for ensuring that the poorest and most vulnerable populations are not excluded from these protections (UNFCCC, 2015).

However, to effectively address the risks posed by climate change, insurance premiums must be adjusted to reflect the true risks that climate change brings. As the frequency and severity of extreme weather events increase, insurance premiums will rise, but it is essential that these premiums remain affordable, particularly for vulnerable populations. Subsidies or public-private partnerships may be necessary to ensure that insurance remains accessible and equitable. Additionally, the insurance industry must invest in better risk modeling and data collection to better understand the potential impacts of climate change and improve the design of insurance products that accurately reflect these risks.

Furthermore, risk management strategies that go beyond insurance, such as building resilient infrastructure, implementing early warning systems, and fostering community preparedness, are essential to reducing vulnerability to climate-related disasters. Insurance alone cannot prevent the damage caused by extreme weather, but it can provide a necessary financial buffer while communities implement strategies to reduce risk in the first place. By integrating insurance and risk management strategies with broader adaptation efforts, governments, businesses, and individuals can enhance their resilience to the growing impacts of climate change.

The Role of Global Cooperation

The role of global cooperation in addressing climate change cannot be overstated. Climate change is a complex, interconnected challenge that transcends national borders, requiring collective action on a global scale. No single country can solve this crisis alone, as the impacts of climate change, whether through rising sea levels, extreme weather events, or shifting agricultural patterns, affect every corner of the globe. Global cooperation is essential for sharing knowledge, resources, and technology to mitigate emissions, support adaptation efforts, and ensure a sustainable future for all. International frameworks, such as the Paris Agreement, play a pivotal role in aligning the efforts of countries and establishing commitments to reduce greenhouse gas emissions. Strengthening these agreements, fostering innovation, and increasing financial support for vulnerable nations are all necessary components of a global response to climate change. Through collaborative efforts, countries can leverage their collective strengths, address disparities, and create an equitable path forward in the fight against climate change.

Strengthening International Agreements

One of the most critical components in addressing climate change effectively is the strengthening of international agreements, such as the Paris Agreement. The Paris Agreement, signed in 2015, was a landmark achievement in global climate diplomacy, with nearly every nation committing to limit global temperature rise to well below 2°C, with an aspiration to limit it to 1.5°C above pre-industrial levels. This ambitious target aims to prevent the most catastrophic impacts of climate change, such as extreme weather events, rising sea levels, and the widespread loss of biodiversity. However, the Paris Agreement's success hinges on countries enhancing their commitments over time and aligning their national policies with global climate goals.

High-emission nations, particularly those that have historically contributed the most to greenhouse gas emissions, must take a leading role in this effort. These countries are responsible for a significant portion of the emissions that have already caused global warming, and they have a moral and historical responsibility to lead the way in reducing emissions. This leadership should not only focus on national reductions but also on supporting developing nations, which have contributed far less to the problem but are often the most vulnerable to climate impacts. Developed countries must fulfill their obligations by providing financial resources, transferring clean technologies, and building capacity to help developing countries mitigate their emissions and adapt to the impacts of climate change.

Achieving the goals of the Paris Agreement will require continued global cooperation and the implementation of more ambitious emission reduction targets by all parties. Regularly updating Nationally Determined Contributions (NDCs) and ensuring that countries meet their emission reduction commitments will be crucial to keeping global warming below 1.5°C. Only through global collaboration, where nations hold each other accountable, can the necessary scale of action be achieved to mitigate the impacts of climate change and avoid the most severe consequences for the planet and its inhabitants (UNFCCC, 2015).

Shared Innovation and Knowledge

Collaboration between countries, industries, and research institutions is indispensable to accelerating the development and deployment of renewable technologies, climate science, and sustainable practices. Climate change is a global issue that requires a global solution, and sharing knowledge and best practices can help countries overcome challenges and accelerate progress toward a more sustainable future. By pooling resources and expertise, nations can address the technological, financial, and logistical barriers that hinder the widespread adoption of renewable energy, energy efficiency measures, and climate adaptation strategies.

Public-private partnerships are vital to driving innovation in renewable energy and low-carbon technologies. Governments, with their ability to set policy frameworks and provide financial incentives, must work with the private sector, which brings innovation, investment, and expertise to the table. International research collaborations, such as the International Energy Agency (IEA) or the Intergovernmental Panel on Climate Change (IPCC), have been essential in advancing climate science and developing solutions that are critical to understanding and mitigating the effects of climate change. These platforms enable countries to share data, monitor emissions, and track progress in real-time, ensuring that the global community is moving toward common goals.

Cross-sector innovation is also a key driver of the transition to a sustainable global economy. Cooperation between sectors such as energy, agriculture, transportation, and water management can lead to integrated solutions that reduce emissions while improving efficiency and resilience. For example, innovations in the agricultural sector, such as precision farming and vertical farming, can help reduce emissions and increase food security, while advancements in electric transportation and energy storage are transforming how societies approach mobility and energy consumption. These integrated, cross-sector strategies will help foster the transition to a green, low-carbon economy, creating new industries, jobs, and opportunities while reducing the environmental footprint.

Funding for Adaptation and Mitigation

In addition to enhancing international agreements and fostering innovation, increasing financial support for adaptation and mitigation efforts is essential to building resilience in vulnerable regions. Climate change disproportionately affects developing countries, which often lack the financial resources and infrastructure to adapt to the changing climate. Despite contributing the least to global greenhouse gas emissions, these countries are bearing the brunt of the consequences, whether through more severe droughts, rising sea levels, or extreme weather events. Therefore, developed nations have both a moral and financial responsibility to provide support for these countries' efforts to mitigate and adapt to climate change.

Climate financing initiatives, such as the Green Climate Fund (GCF), play a central role in addressing the financial needs of developing countries. The GCF, established under the United Nations Framework Convention on Climate Change (UNFCCC), is designed to provide financial assistance to developing countries to implement projects that reduce emissions (mitigation) and build resilience to climate impacts (adaptation). This fund is vital to helping low-income countries access the resources needed to adopt sustainable energy solutions, enhance infrastructure resilience, improve agricultural practices, and protect vulnerable communities from the adverse effects of climate change (UNFCCC, 2015). However, despite progress, there is still a significant gap between the financing needed and the funds available, particularly for the most vulnerable countries.

Strengthening and expanding these funding mechanisms will be key to ensuring that developing countries can effectively address climate change. In addition to the GCF, innovative financing models, such as green bonds, climate insurance, and climate risk financing, can provide additional resources for adaptation and mitigation. It is also essential that financial support be coupled with capacity-building efforts that ensure countries have the knowledge, skills, and governance structures needed to implement climate projects effectively. Furthermore, ensuring that funds are used efficiently and transparently will be crucial for building trust and ensuring that financial resources are used to maximize the impacts of adaptation and mitigation projects.

By increasing the financial flows to vulnerable regions, the global community can help ensure that these nations are not left behind in the fight against climate change. Investing in adaptation and mitigation will not only reduce the future costs of climate change but will also create opportunities for sustainable development, poverty alleviation, and climate resilience in some of the world's most vulnerable regions.

Empowering Communities and Individuals

Empowering communities and individuals is a fundamental aspect of tackling climate change. While large-scale systemic changes are necessary to address the global nature of the crisis, the actions of local communities and individuals can significantly contribute to mitigating climate change and building resilience. By fostering grassroots initiatives, encouraging sustainable behaviors, and providing access to education and resources, we can equip people with the tools they need to take meaningful action. Whether through community-based renewable energy projects, sustainable agriculture practices, or changes in consumption habits, individuals and local organizations have the potential to drive change from the ground up. Empowering people not only enhances the collective response to climate change but also ensures that adaptation and mitigation strategies are equitable, inclusive, and responsive to local needs.

Local Solutions to Global Problems

While international cooperation and large-scale policy initiatives are critical in addressing climate change, local solutions are equally important. Grassroots initiatives and community-based actions can significantly contribute to reducing emissions, increasing resilience, and fostering long-term sustainability. Local solutions are often more flexible and adaptable, tailoring strategies to specific regional needs, cultures, and circumstances. When scaled up, these actions can create powerful movements that contribute to achieving global climate goals.

One example of a local solution is community-based renewable energy projects. In many areas, particularly in rural and underserved regions, communities have taken the initiative to build solar or wind energy systems, reducing their dependence on fossil fuels and increasing energy security. These projects not only provide clean energy but also create jobs and foster local economic development. Community-driven renewable energy systems, such as solar cooperatives, can empower residents to take control of their energy supply, lower energy costs, and reduce their carbon footprints. In addition, local conservation efforts, such as the restoration of degraded ecosystems, protecting local wildlife habitats, and enhancing biodiversity, can help mitigate climate impacts while fostering resilience to changing weather patterns.

Urban farming is another effective local solution that can address both food security and environmental sustainability. By transforming unused urban spaces into productive food gardens, cities can reduce the carbon footprint of food production, shorten supply chains, and promote sustainable agricultural practices. Urban farming projects also contribute to local food sovereignty, reduce food waste, and improve community health by providing access to fresh, nutritious produce. These initiatives demonstrate how local action can help mitigate climate change and provide valuable lessons that can be shared with other regions.

When these local solutions are scaled up and integrated into national and global climate strategies, they become powerful tools for reducing emissions and enhancing resilience at the community level. By promoting and supporting grassroots initiatives, governments and organizations can create an environment where local solutions thrive and contribute meaningfully to the fight against climate change.

Behavioral Changes

Individual actions, though often overlooked in large-scale discussions about climate change, play a crucial role in reducing emissions and fostering a sustainable future. While structural and policy changes are essential, shifting the behavior of individuals is equally important in achieving lasting change. Everyday decisions, from the food we eat to the products we purchase, have significant environmental impacts.

One of the most effective behavioral changes individuals can make is adjusting their diets. Reducing meat consumption, particularly beef and other ruminant livestock, can substantially lower the carbon footprint of food production. Livestock farming is a major contributor to greenhouse gas emissions, particularly methane, which has a much higher global warming potential than carbon dioxide (Smith et al., 2014). By embracing plant-based diets or reducing meat intake, individuals can significantly reduce their environmental impact. Additionally, supporting sustainable food production practices, such as purchasing locally grown and organic foods, can reduce transportation emissions and promote environmentally friendly agricultural practices.

Beyond dietary changes, individuals can adopt more sustainable lifestyles in many other areas. Reducing waste by recycling, composting, and reusing materials helps prevent landfill overflow and reduces methane emissions. Conserving energy in daily life—by using energy-efficient appliances, turning off lights when not in use, and utilizing public transportation, also contributes to lowering personal carbon footprints. Moreover, choosing sustainable products, such as those made from recycled materials or with minimal packaging, can reduce environmental impact at every stage of production, from manufacturing to disposal.

While individual actions alone will not solve the climate crisis, they are a critical part of the broader solution. When combined with policy-driven initiatives, technological innovations, and large-scale structural changes, individual behavioral shifts can make a meaningful contribution. By collectively adopting more sustainable habits, individuals can help create a culture of climate-conscious living that supports wider systemic changes.

Education and Awareness

Empowering future generations with the knowledge and tools to tackle climate change is essential for long-term success. Education plays a pivotal role in shaping attitudes, behaviors, and the capacity to address climate challenges. The more individuals understand about climate change, its impacts, and the available solutions, the more likely they are to make informed choices and advocate for effective climate policies.

Integrating climate education into school curricula is a critical step toward building a generation that is well-equipped to confront climate challenges. By teaching students about the science of climate change, the importance of sustainability, and the need for immediate action, schools can foster a sense of environmental stewardship that will extend into adulthood. Climate education can also empower young people to become leaders in the fight against climate change, equipping them with the knowledge to make impactful decisions in their careers and communities. This education should also emphasize the interconnectedness of environmental, social, and economic issues, helping students understand the broader implications of climate change on global inequality, public health, and human rights.

In addition to formal education, fostering awareness through media campaigns, public discussions, and community outreach is crucial to ensuring that climate change remains a priority at every level of society. Public awareness campaigns can inspire individuals to take action in their own lives, encouraging them to adopt more sustainable behaviors and support climate policies. Additionally, these efforts can create a groundswell of demand for stronger climate action, influencing political leaders and policymakers to act decisively.

Ultimately, education and awareness efforts must be comprehensive, inclusive, and global. They should equip individuals with the understanding, tools, and motivation to tackle climate change in their own lives while encouraging broader societal change. By fostering a climate-literate society, we can ensure that future generations are prepared to face the challenges of a changing climate and work toward a sustainable and resilient future.

Sustainable Coexistence with Climate Change

Finally, in the face of an undeniable climate crisis, it is clear that halting climate change entirely is no longer within our grasp. However, what is within our power is the ability to reduce our impact, adapt to the changes already set in motion, and build the resilience needed to thrive in a shifting world. With the right tools, knowledge, and determination, we can mitigate the most devastating effects of climate change while learning to coexist with a planet in flux. The journey ahead requires us to come together, innovate, and remain steadfast in our commitment to securing a livable world for future generations. By embracing this collective responsibility, we can not only navigate the complex challenges of climate change but also pave the way toward a sustainable, equitable future where both people and the planet can flourish. The time to act is now, and the opportunity to create meaningful change is in our hands.

Chapter 17

Earth's Last Warning

Earth's climate has always been a story of change, marked by cycles of warming and cooling that unfolded over millennia. For billions of years, this intricate system has been shaped by natural forces, providing the backdrop for life to evolve and thrive. Yet today, we face an unprecedented challenge: a rapidly shifting climate driven not by natural forces, but by human activity, altering the trajectory of this ancient narrative at breakneck speed.

Recap of Earth's Dynamic Climate Journey

Earth's climate has been in a state of constant flux for billions of years, shaped by complex interactions between the atmosphere, oceans, landmasses, and the organisms that inhabit the planet. From the early days of the Earth, when the planet was a hot, molten mass, to the formation of a more stable climate system, Earth's climate has oscillated between periods of intense warmth and extreme cold. These shifts were influenced by natural factors such as volcanic activity, the Earth's orbital variations, and the gradual changes in solar output. However, the current climate crisis is unique, as the rapid pace of change today is driven largely by human activity.

Over the last several million years, Earth has experienced cyclical glaciations and interglacial periods, with significant ice sheets forming and receding. During the last Ice Age, which peaked around 20,000 years ago, vast ice sheets covered large portions of the planet. As the Earth began to warm, these ice sheets retreated, giving way to the relatively stable climate conditions that we have experienced in the past 10,000 years, known as the Holocene epoch. This period of relative climate stability has allowed human civilization to flourish, but it has also been a time of unprecedented environmental change due to human influence, particularly in the last few centuries.

The Industrial Revolution, beginning in the late 18th century, marked the start of dramatic changes to Earth's climate system. The widespread use of fossil fuels for industry, transportation, and agriculture led to a sharp increase in the concentration of greenhouse gases in the atmosphere, particularly carbon dioxide (CO_2) and methane (CH_4). This increase in greenhouse gases began to drive global temperatures higher, setting in motion the warming we are experiencing today. Over the past century, Earth's average temperature has risen by approximately 1.1°C, with the past few decades seeing an acceleration of this trend (IPCC, 2021). The human influence on the climate is no longer a theory or an isolated phenomenon, it is a stark reality that we are witnessing in real time.

Interconnectedness of Past, Present, and Future Climates

The climate of today is inextricably linked to the climate of the past, and the actions we take today will shape the climate of the future. Earth's climate system is deeply interconnected, with changes in one area often influencing others in complex ways. For instance, rising global temperatures are causing ice sheets and glaciers to melt, which in turn contributes to rising sea levels. The warming of the oceans is altering weather patterns, leading to more frequent and intense hurricanes, droughts, and heatwaves. These changes are not happening in isolation but are interconnected with every aspect of the climate system.

Understanding the past is critical to understanding where we are today. Over geological timescales, Earth's climate has fluctuated, but the current rate of warming is unprecedented. Human activities, particularly the burning of fossil fuels and deforestation, have tipped the balance of Earth's climate system, introducing changes at a pace that ecosystems and human societies are struggling to adapt to. The carbon emissions that we continue to release into the atmosphere today will affect the climate for centuries, and the actions we take now will determine the trajectory of future climate conditions. The past has set the stage for the present, and the present will, in turn, shape the future.

The urgency of addressing climate change is clear: we are at a critical juncture in history. The choices made today will determine the extent of future warming, the severity of climate-related impacts, and the ability of future generations to adapt. There is no denying that the path we are on is unsustainable, but understanding the interconnectedness of the past, present, and future climates can help guide our decisions and actions in the coming decades.

Can Humans Stop, Reverse, or Repair the Climate?

This question is perhaps one of the most pressing and difficult to answer. The reality is that Earth's climate has already undergone significant changes that cannot be undone. Even if we stopped emitting greenhouse gases tomorrow, the effects of past emissions would continue to affect the climate for many decades to come. The carbon dioxide and methane that have already been released into the atmosphere will continue to trap heat, leading to further warming and disruption of climate systems. However, this does not mean that all hope is lost or that efforts to mitigate climate change are futile.

There is still time to slow the rate of warming, reduce its severity, and avoid the worst impacts of climate change. Scientists agree that limiting global temperature rise to well below 2°C, with efforts to keep it under 1.5°C, is critical to avoid catastrophic climate impacts (IPCC, 2021).

Achieving this goal will require unprecedented global cooperation, massive reductions in greenhouse gas emissions, and transformative changes to how we produce and consume energy, food, and materials. But even if we do not meet these ambitious targets, it is still possible to reduce the severity of the impacts and build resilience to the changes that are already underway.

Humans have the knowledge, technology, and resources to significantly reduce emissions and adapt to the changing climate. Renewable energy technologies, such as wind, solar, and hydropower, are increasingly cost-competitive and can provide the clean energy needed to reduce reliance on fossil fuels. Energy efficiency measures, changes in land-use practices, and carbon capture technologies also offer pathways to reduce the amount of CO_2 in the atmosphere. Moreover, efforts to conserve and restore ecosystems, such as reforestation and wetland restoration, can help sequester carbon and improve resilience to climate impacts.

However, even with these solutions, we cannot "repair" the climate in the traditional sense. The idea of returning to a pre-industrial climate is unrealistic. What is possible is the mitigation of further damage and the adaptation to a new climate reality. By taking immediate action to reduce emissions, increase carbon sequestration, and build resilient infrastructure, we can minimize the worst consequences of climate change. The future is still within our control, but only if we act swiftly, boldly, and with a sense of collective responsibility.

Encouragement to Use Historical Insights for Global Climate Responsibility

As we face the challenges of climate change, it is essential to look to history for guidance. The Earth's climate story is not one of linear progress but a cyclical tale of change, adaptation, and, at times, collapse. From the rise and fall of ancient civilizations to the dramatic environmental shifts that have redrawn the map of human destiny, history teaches us two crucial truths: the planet is always changing, and those who refuse to change with it are often left behind.

Time and again, civilizations have faltered not because they lacked intelligence, strength, or vision, but because they failed to respond when the climate shifted beneath their feet. The Ancestral Puebloans, the Maya, the Akkadians, and the Norse Greenlanders all reached tipping points, when fragile systems cracked under pressure, and resilience gave way to ruin. Their stories are not relics of a distant past. They are warning signs etched into the ruins, whispering of what happens when we delay action, deny reality, or cling too tightly to what no longer serves us.

And now, the cycle has come full circle. But this time, we are not just part of the story, we are writing it. We are no longer reacting to nature's shifts; we are driving them. Our industries, our choices, and our appetite for growth have become forces of geological change. The question that tomorrow's archaeologists may ask is not whether we knew what was happening, but why, with all our knowledge, we chose not to act.

Yet, within this moment lies our greatest opportunity. History also teaches us that humanity is capable of remarkable things when we unite in common purpose. One critical historical insight is the importance of global cooperation. Climate change, like many of the world's most significant challenges, is a global problem that demands a global response. The Paris Agreement, while a crucial step, is only the beginning. History has shown that when nations come together, like they did to confront the ozone crisis in the 1980s, meaningful change is possible (United Nations, 1985). We must summon that same will, now.

Innovation, too, is one of our greatest weapons. Throughout history, we have faced plagues, wars, and disasters, and emerged more adaptable and ingenious than before. Today, our scientific and technological capacity far exceeds that of any previous age. We must now harness that power not just to survive climate change, but to transform it into the catalyst for a better, more resilient world.

Ultimately, the key lesson from Earth's climate story is this: the fate of the planet is not predetermined, it's in our hands. While the scale of the threat is immense, so too is the potential for transformation. By learning from the past, embracing innovation, and committing to bold, collective action, we can choose not to be remembered as the civilization that let it all slip away, but as the one that turned back from the edge and built something enduring.

The challenge is great. But the opportunity? Even greater. Let future generations look back, not in sorrow at what was lost, but in gratitude for what we chose to save.

Final thought of our beautiful blue planet and its Climate Crisis:

> "The environment is improving in many ways, and the key to solving the rest of the problems is to find solutions that are more efficient and effective, not to abandon progress altogether."
>
> -   Bjorn Lomborg

Bibliography

Alley, R. B. (2000). The Younger Dryas cold interval as viewed from central Greenland. Quaternary Science Reviews, 19(1–5), 213–226. https://doi.org/10.1016/S0277-3791(99)00062-1

Arneborg, J., Heinemeier, J., Lynnerup, N., Nielsen, H. L., Rud, N., & Sveinbjörnsdóttir, Á. E. (1999). Change of diet of the Greenland Vikings determined from stable carbon isotope analysis and 14C dating of their bones. Radiocarbon, 41(2), 157–168. https://doi.org/10.1017/S0033822200019512

Andrew, R. M. (2018). Global carbon intensity of cement production. Environmental Science & Technology, 52(2), 1274-1281.

Armstrong, P. H. (2009). Landscapes and identity in early medieval Britain. Cambridge University Press.

Arrhenius, S. (1896). On the influence of carbonic acid in the air upon the temperature of the ground. Philosophical Magazine, 41(251), 237-276. https://doi.org/10.1080/14786449608620762

Behringer, W. (1999). Climatic change and witch-hunting: The impact of the Little Ice Age on mentalities. Climatic Change, 43(1), 335–351. https://doi.org/10.1023/A:1005554519604

Behringer, W. (1999). Climatic change and witch-hunting: The impact of the Little Ice Age on mentalities. Climatic Change, 43(1), 335–351. https://doi.org/10.1023/A:1005554519604

Benn, D. I., & Evans, D. J. A. (2010). Glaciers and glaciation (2nd ed.). Routledge.

Benson, L. V., Berry, M. S., Jolie, E. A., Spangler, J. D., Stahle, D. W., & Hattori, E. M. (2007). Possible impacts of early-11th-, middle-12th-, and late-13th-century droughts on western Native Americans and the Mississippian Cahokians. Quaternary Science

Reviews, 26(3–4), 336–350.
https://doi.org/10.1016/j.quascirev.2006.08.020

Berger, A., & Loutre, M. F. (2003). Climate 400,000 years ago, a key to the future? Geophysical Monograph Series, 137, 17–26. https://doi.org/10.1029/137GM02

Berner, R. A. (2004). The Phanerozoic carbon cycle: CO_2 and O_2. Oxford University Press.

Broecker, W. S. (1991). The great ocean conveyor. Oceanography, 4(2), 79-89.

Broecker, W. S. (1997). Thermohaline circulation, the Achilles heel of our climate system: Will man-made CO_2 upset the current balance? Science, 278(5343), 1582–1588. https://doi.org/10.1126/science.278.5343.1582

Broecker, W. S. (2006). Was the Younger Dryas triggered by a flood? Science, 312(5777), 1146–1148. https://doi.org/10.1126/science.1123253

Campbell, B. M. S. (2016). The Great Transition: Climate, Disease and Society in the Late-Medieval World. Cambridge University Press.

Carlson, A. E. (2010). What caused the Younger Dryas cold event? Geology, 38(4), 383–384. https://doi.org/10.1130/focus042010.1

Ceballos, G., Ehrlich, P. R., & Dirzo, R. (2015). Biological annihilation via the ongoing sixth mass extinction signaled by vertebrate population losses and declines. Proceedings of the National Academy of Sciences, 113(29), 8079-8086. https://doi.org/10.1073/pnas.1521536113

Clague, J. J., & Ward, B. (2011). Pleistocene glaciation of British Columbia. Developments in Quaternary Science, 15, 563–573. https://doi.org/10.1016/B978-0-444-53447-7.00044-1

Clapperton, C. M. (1993). Quaternary geology and geomorphology of South America. Elsevier Science. https://doi.org/10.1016/B978-0-444-89897-3.50006-7

Clark, P. U., Dyke, A. S., Shakun, J. D., Carlson, A. E., Clark, J., Wohlfarth, B., ... & McCabe, A. M. (2009). The Last Glacial Maximum. Science, 325(5941), 710-714. https://doi.org/10.1126/science.1172873

Clark, C. D., Hughes, A. L. C., Greenwood, S. L., Jordan, C., & Sejrup, H. P. (2012). Pattern and timing of retreat of the last British-Irish Ice Sheet. Quaternary Science Reviews, 44, 112-146. https://doi.org/10.1016/j.quascirev.2010.07.019

Clayton, S. (2020). The psychology of climate change communication. Cambridge University Press.

Clottes, J. (2008). Cave art. Phaidon Press.

Coles, B. (2002). Becoming a wetland archaeologist: A personal perspective. The Antiquaries Journal, 82, 51-68. https://doi.org/10.1017/S0003581500070345

Cullen, H. M., deMenocal, P. B., Hemming, S., Hemming, G., Brown, F. H., Guilderson, T., & Sirocko, F. (2000). Climate change and the collapse of the Akkadian Empire: Evidence from the deep sea. Geology, 28(4), 379–382. https://doi.org/10.1130/0091-7613(2000)28<379:CCATCO>2.0.CO;2

Crutzen, P. J. (2002). Geology of mankind. Nature, 415(6867), 23. https://doi.org/10.1038/415023a

Dean, J. S., & Funkhouser, G. S. (2002). Dendroclimatic reconstructions from the southwestern United States. In R. S.

Bradley & P. D. Jones (Eds.), Climate Since A.D. 1500 (pp. 163–184). Routledge.

Diamond, J. (2005). Collapse: How societies choose to fail or succeed. Viking Press.

Dickens, G. R. (2011). Down the rabbit hole: Toward appropriate discussion of methane release from gas hydrate systems during the Paleocene-Eocene Thermal Maximum and other past hyperthermal events. Climatic Change, 107(3–4), 267–276. https://doi.org/10.1007/s10584-011-0072-y

Douglas, P. M. J., Demarest, A. A., Brenner, M., & Canuto, M. A. (2015). Impacts of climate change on the collapse of lowland Maya civilization. Annual Review of Earth and Planetary Sciences, 44, 613–645. https://doi.org/10.1146/annurev-earth-060115-012512

Dugmore, A. J., Keller, C., & McGovern, T. H. (2007). Norse Greenland settlement: Reflections on climate change, trade, and the contrasting fates of human settlements in the North Atlantic islands. Arctic Anthropology, 44(1), 12–36. https://doi.org/10.1353/arc.2011.0038

Dutton, A., & Lambeck, K. (2012). Ice volume and sea level during the last interglacial. Science, 337(6091), 216–219. https://doi.org/10.1126/science.1205749

Dyke, A. S., Moore, A., & Robertson, L. (2002). Deglaciation of North America. Geological Survey of Canada Open File, 1574, 1–27.

Eddy, J. A. (1976). The Maunder Minimum. Science, 192(4245), 1189–1202. https://doi.org/10.1126/science.192.4245.1189

Ehlers, J., & Gibbard, P. L. (2007). The extent and chronology of Cenozoic Global Glaciation. Quaternary International, 164(1), 6-20. https://doi.org/10.1016/j.quaint.2006.09.019

Ehlers, J., Gibbard, P. L., & Hughes, P. D. (Eds.). (2011). Developments in Quaternary Science: Quaternary glaciations - extent and chronology. Elsevier.

Esper, J., Cook, E. R., & Schweingruber, F. H. (2002). Low-frequency signals in long tree-ring chronologies for reconstructing past temperature variability. Science, 295(5563), 2250–2253. https://doi.org/10.1126/science.1066208

Eyles, N. (2012). Glacial geology: A landsystems approach. Hodder Education.

Fagan, B. (2011). Cro-Magnon: How the Ice Age gave birth to the first modern humans. Bloomsbury Press.

FAO. (2018). The state of food and agriculture: Migration, agriculture, and rural development. Food and Agriculture Organization of the United Nations.

Fairchild, I. J., & Baker, A. (2012). Speleothem science: From process to past environments. Wiley-Blackwell.

Faith, J. T., & Surovell, T. A. (2009). Synchronous extinction of North America's Pleistocene mammals. Proceedings of the National Academy of Sciences, 106(49), 20641–20645. https://doi.org/10.1073/pnas.0908153106

Fourier, J. (1824). Mémoire sur les températures du globe terrestre et des espaces planétaires. Mémoires de l'Académie Royale des Sciences, 7, 571-604.

Gaffney, V., Fitch, S., & Smith, D. (2009). Europe's lost world: The rediscovery of Doggerland. Council for British Archaeology.

Gifford, R. (2011). The dragons of inaction: Psychological barriers that limit climate change mitigation and adaptation. American Psychologist, 66(4), 290–302. https://doi.org/10.1037/a0023566

Goebel, T., Waters, M. R., & O'Rourke, D. H. (2008). The Late Pleistocene dispersal of modern humans in the Americas. Science, 319(5869), 1497–1502. https://doi.org/10.1126/science.1153569

Grove, J. M. (2004). The Little Ice Age. Routledge.

Hansen, J., et al. (1981). Climate impact of increasing atmospheric carbon dioxide. Science, 213(4511), 957-966. https://doi.org/10.1126/science.213.4511.957

Hawken, P. (2017). Drawdown: The most comprehensive plan ever proposed to reverse global warming. Penguin.

Hays, J. D., Imbrie, J., & Shackleton, N. J. (1976). Variations in the Earth's orbit: Pacemaker of the ice ages. Science, 194(4270), 1121-1132. https://doi.org/10.1126/science.194.4270.1121

Heusser, C. J. (2003). Ice age Southern Andes: A chronicle of paleoecological events. Elsevier. https://doi.org/10.1016/B978-0-444-51540-5.X5000-0

Hodell, D. A., Curtis, J. H., & Brenner, M. (1995). Possible role of climate in the collapse of Classic Maya civilization. Nature, 375(6530), 391–394. https://doi.org/10.1038/375391a0

Hoegh-Guldberg, O., Poloczanska, E. S., Skirving, W., & Dove, S. (2018). Coral reef ecosystems under climate change and ocean acidification. Frontiers in Marine Science, 4, 158. https://doi.org/10.3389/fmars.2017.00158

Houghton, R. A. (2015). Tropical deforestation as a source of greenhouse gas emissions. Environmental Science & Policy, 55, 60-65.

Hulton, N. R. J., Sugden, D. E., Payne, A. J., & Clapperton, C. M. (2002). The Patagonian Ice Sheet during the Last Glacial Maximum and its subsequent decay pattern: Evidence from numerical modeling. Quaternary Research, 57(2), 191–199. https://doi.org/10.1006/qres.2001.2312

IEA. (2020). World Energy Outlook 2020. International Energy Agency.

International Renewable Energy Agency (IRENA). (2020). Renewable power generation costs in 2019. International Renewable Energy Agency. https://www.irena.org/Publications/2020/Jun/Renewable-power-generation-costs-in-2019

Intergovernmental Panel on Climate Change (IPCC). (2021). Climate change 2021: The physical science basis. Contribution of Working Group I to the Sixth Assessment Report of the Intergovernmental Panel on Climate Change. Cambridge University Press. https://doi.org/10.1017/9781009157896

Jones, C. A., & Durbin, E. G. (2018). The role of coal in the global economy. Energy Economics, 72, 1-9.

Keeling, C. D. (1960). The concentration and isotopic abundances of carbon dioxide in the atmosphere. Geochimica et Cosmochimica Acta, 28(5), 473-490. https://doi.org/10.1016/0016-7037(60)90012-4

Keith, D. W., et al. (2018). A process for capturing CO_2 from the atmosphere. Joule, 2(8), 1573-1592. https://doi.org/10.1016/j.joule.2018.05.006

Kirschvink, J. L. (1992). Late Proterozoic low-latitude global glaciation: The snowball Earth. In J. W. Schopf & C. Klein (Eds.), The Proterozoic Biosphere: A Multidisciplinary Study (pp. 51–52). Cambridge University Press.

Lal, R. (2005). Forest soils and carbon sequestration. Forest Ecology and Management, 220(1-3), 7-19. https://doi.org/10.1016/j.foreco.2005.08.001

Lamb, H. H. (1965). The early medieval warm epoch and its sequel. Palaeogeography, Palaeoclimatology, Palaeoecology, 1(1), 13–37. https://doi.org/10.1016/0031-0182(65)90004-0

Lambeck, K., Rouby, H., Purcell, A., Sun, Y., & Sambridge, M. (2014). Sea level and global ice volumes from the Last Glacial Maximum to the Holocene. Proceedings of the National Academy of Sciences, 111(43), 15296-15303. https://doi.org/10.1073/pnas.1411762111

Lashof, D. A., & Ahuja, D. (1990). Relative contributions of greenhouse gas emissions to global warming. Nature, 344(6266), 529-531. https://doi.org/10.1038/344529a0

Lekson, S. H. (2006). The Chaco Meridian: Centers of political power in the ancient Southwest (2nd ed.). AltaMira Press.

Lenton, T. M., et al. (2019). Climate tipping points — too risky to bet against. Nature, 575(7784), 592-595.

Lindsey, R. (2021). Climate Change: Global Temperature. National Oceanic and Atmospheric Administration. https://www.climate.gov

Lomborg, B. (2001). The skeptical environmentalist: Measuring the real state of the world. Cambridge University Press.

Lucero, L. J. (2002). The collapse of the Classic Maya: A case for the role of water control. American Anthropologist, 104(3), 814–826. https://doi.org/10.1525/aa.2002.104.3.814

Lund, H. (2020). Renewable energy systems: A smart energy systems approach to the future of the grid. Academic Press.

Mann, M. E., Zhang, Z., Rutherford, S., Bradley, R. S., Hughes, M. K., Shindell, D., ... & Ammann, C. (2009). Global signatures and dynamical origins of the Little Ice Age and Medieval Climate Anomaly. Science, 326(5957), 1256–1260. https://doi.org/10.1126/science.1177303

Marlow, D. R., Maharaj, N., & Barker, M. (2021). Managing Day Zero: How Cape Town avoided running out of water. Water Policy, 23(4), 765–780. https://doi.org/10.2166/wp.2021.105

Martin, S., & Grube, N. (2008). Chronicle of the Maya Kings and Queens: Deciphering the Dynasties of the Ancient Maya (2nd ed.). Thames & Hudson.

McGovern, T. H. (1990). The demise of Norse Greenland. In G. L. Webster (Ed.), Archaeological thought in America (pp. 204–213). Cambridge University Press.

McInerney, F. A., & Wing, S. L. (2011). The Paleocene-Eocene Thermal Maximum: A perturbation of carbon cycle, climate, and biosphere with implications for the future. Annual Review of Earth and Planetary Sciences, 39, 489–516. https://doi.org/10.1146/annurev-earth-040610-133431

McKibben, B. (1989). The end of nature. Random House.

Milankovitch, M. (1941). Kanon der Erdbestrahlung und seine Anwendung auf das Eiszeitenproblem. Royal Serbian Academy.

Miller, J. R. (2019). Restoring wetlands for carbon sequestration and ecological resilience. Nature Sustainability, 2, 1-8. https://doi.org/10.1038/s41893-019-0241-2

NASA. (2020). Arctic sea ice minimum shows 2nd lowest extent on record. NASA. https://www.nasa.gov

National Oceanic and Atmospheric Administration (NOAA). (2020). State of the Climate: Hurricanes and Tropical Storms. NOAA.

National Oceanic and Atmospheric Administration (NOAA). (2023). Trends in atmospheric carbon dioxide. Retrieved from https://www.noaa.gov

NOAA. (2023). Trends in atmospheric carbon dioxide. National Oceanic and Atmospheric Administration. https://gml.noaa.gov/ccgg/trends/

Ogilvie, A. E. J., & Jónsson, T. (2000). "Little Ice Age" research: A perspective from Iceland. Climatic Change, 48(1), 9–52. https://doi.org/10.1023/A:1005663729564

Ogilvie, A. E. J., & Jónsson, T. (2001). "Little Ice Age" research: A perspective from Iceland. Climatic Change, 48(1), 9–52. https://doi.org/10.1023/A:1005625729889

Oppenheimer, C. (2003). Climatic, environmental and human consequences of the largest known historic eruption: Tambora volcano (Indonesia) 1815. Progress in Physical Geography, 27(2), 230–259. https://doi.org/10.1191/0309133303pp379ra

Perri, A. R. (2019). A wolf in dog's clothing: Initial dog domestication and Pleistocene human society. Journal of Archaeological Science, 92, 1-8. https://doi.org/10.1016/j.jas.2018.12.002

Petit, J. R., Jouzel, J., Raynaud, D., Barkov, N. I., Barnola, J. M., Basile, I., ... & Stievenard, M. (1999). Climate and atmospheric history of the past 420,000 years from the Vostok ice core, Antarctica. Nature, 399(6735), 429-436. https://doi.org/10.1038/20859

Ponting, C. (2007). A new green history of the world: The environment and the collapse of great civilizations. Penguin Books.

Rabassa, J., & Clapperton, C. M. (1990). Quaternary glaciations in the Southern Hemisphere: An overview. Quaternary Science

Reviews, 9(2), 299–304. https://doi.org/10.1016/0277-3791(90)90017-Q

Rahmstorf, S., Archer, D., Eby, M., Brovkin, V., Ganopolski, A., Lenton, T. M., ... & Zickfeld, K. (2005). The thermohaline circulation and its response to abrupt climate change. Quaternary Science Reviews, 24(10–11), 1399–1410. https://doi.org/10.1016/j.quascirev.2005.04.005

Rockström, J., et al. (2009). A safe operating space for humanity. Nature, 461(7263), 472-475.

Roberts, B. W. (2014). The age of agriculture: From foraging to farming. Routledge.

Royer, D. L. (2014). Atmospheric CO_2 and O_2 during the Phanerozoic: Tools, patterns, and impacts. Treatise on Geochemistry, 6, 251–267. https://doi.org/10.1016/B978-0-08-095975-7.01311-5

Ruddiman, W. F. (2014). Earth's Climate: Past and Future. W.H. Freeman and Company.

Schuur, E. A. G., et al. (2015). Climatic change and the permafrost-carbon feedback. Nature, 520(7546), 171-179.

Screen, J. A., & Simmonds, I. (2010). The central role of the southern annular mode in regional climate variability. Nature Geoscience, 3(9), 743-749.

Serreze, M. C., & Barry, R. G. (2011). Processes and impacts of Arctic amplification: A research synthesis. Global and Planetary Change, 77(1-2), 85-96. https://doi.org/10.1016/j.gloplacha.2011.03.004

Smith, P., et al. (2014). Agriculture, forestry and other land use (AFOLU). In Climate Change 2014: Mitigation of Climate Change (pp. 811-922). Cambridge University Press.

Steffen, W., Crutzen, P. J., & McNeill, J. R. (2011). The Anthropocene: Are humans now overwhelming the great forces of nature? Ambio, 36(8), 614–621. https://doi.org/10.1579/0044-7447

Steffen, W., et al. (2015). The trajectory of the Anthropocene: The Great Acceleration. The Anthropocene Review, 2(1), 81-98. https://doi.org/10.1177/2053019614564785

Stern, N. (2007). The economics of climate change: The Stern review. Cambridge University Press.

Sugden, D. E., Bentley, M. J., & O'Cofaigh, C. (2005). Geological and geomorphological insights into Antarctic glaciation and ice sheet dynamics. Quaternary Science Reviews, 24(11–12), 1471–1486. https://doi.org/10.1016/j.quascirev.2005.03.011

Steffen, W., Rockström, J., Richardson, K., Lenton, T. M., Folke, C., Liverman, D., ... & Schellnhuber, H. J. (2018). Trajectories of the Earth system in the Anthropocene. Proceedings of the National Academy of Sciences, 115(33), 8252–8259. https://doi.org/10.1073/pnas.1810141115

Teller, J. T., Leverington, D. W., & Mann, J. D. (2002). Freshwater outbursts to the oceans from glacial Lake Agassiz and their role in climate change. Quaternary Science Reviews, 21(8), 879–887. https://doi.org/10.1016/S0277-3791(01)00145-7

Trenberth, K. E., Fasullo, J. T., & Kiehl, J. (2009). Earth's global energy budget. Bulletin of the American Meteorological Society, 90(3), 311–324. https://doi.org/10.1175/2008BAMS2634.1

Trouet, V., Esper, J., Graham, N. E., Baker, A., Scourse, J. D., & Frank, D. C. (2009). Persistent positive North Atlantic Oscillation mode dominated the Medieval Climate Anomaly. Science, 324(5923), 78–80. https://doi.org/10.1126/science.1166349

Tzedakis, P. C., Emerson, B. C., & Hewitt, G. M. (2013). Cryptic or mystic? Glacial tree refugia in northern Europe. Trends in Ecology & Evolution, 28(12), 696-704. https://doi.org/10.1016/j.tree.2013.09.001

United Nations. (1985). The Montreal Protocol on Substances that Deplete the Ozone Layer. United Nations Environment Programme.

United Nations Environment Programme (UNEP). (2021). Emissions gap report 2021. UNEP.

UNFCCC. (2015). Paris Agreement. United Nations Framework Convention on Climate Change.

Weiss, H., & Bradley, R. S. (2001). What drives societal collapse? Science, 291(5504), 609–610. https://doi.org/10.1126/science.1058775

Weiss, H., Courty, M.-A., Wetterstrom, W., Guichard, F., Senior, L., Meadow, R., & Curnow, A. (1993). The genesis and collapse of third millennium north Mesopotamian civilization. Science, 261(5124), 995–1004. https://doi.org/10.1126/science.261.5124.995

Williams, A. P., Cook, B. I., Smerdon, J. E., Cook, E. R., Abatzoglou, J. T., Bolles, K., … Seager, R. (2020). Large contribution from anthropogenic warming to an emerging North American megadrought. Science, 368(6488), 314–318. https://doi.org/10.1126/science.aaz9600

Zachos, J. C., Dickens, G. R., & Zeebe, R. E. (2008). An early Cenozoic perspective on greenhouse warming and carbon-cycle dynamics. Nature, 451(7176), 279–283. https://doi.org/10.1038/nature06588

Zilhão, J. (2001). Anatomically archaic, behaviorally modern: The last Neanderthals and their destiny. Current Anthropology, 42(2), 203-230. https://doi.org/10.1086/320255

About the Author

Douglas B. Sims, PhD, is a scientist, environmental researcher, and passionate storyteller of Earth's evolving climate. With over three decades of experience in environmental science, Dr. Sims has dedicated his career to uncovering the deep connections between our planet's past and the challenges we face today.

For more than a decade, Dr. Sims has conducted field research on paleolakes in the American Southwest, exploring the dramatic environmental transformations that turned once lush, humid landscapes into the arid deserts we know today. His work delves into how ancient lakes, wetlands, and fertile basins supported rich ecosystems—and how shifting climates reshaped the land, its vegetation, and its human inhabitants.

Drawing on this extensive background, this book reflects Dr. Sims' commitment to making complex scientific concepts accessible and urgent for a broad audience. His writing bridges the gap between deep time and the pressing realities of our warming world, offering readers not just knowledge, but a rallying call for informed action.

When he's not knee-deep in sediment cores or piecing together Earth's climatic story, Dr. Sims mentors the next generation of environmental scientists, speaks on the intersections of science and public policy, and advocates for solutions rooted in both innovation and historical understanding.

www.ingramcontent.com/pod-product-compliance
Lightning Source LLC
Chambersburg PA
CBHW071726120626
46550CB00002B/396